Hartlepool College of FE Library

T07934

D1081078

LEARNER RESOURCE
CENTRE
WITHDRAWN
HARTLEPOOL
COLLEGE OF FE

Construction technology
volume 1

This book is dedicated to my wife for her unstinting efforts in giving construc-tive criticism on the text, as well as typing it.

Construction technology volume 1

J. T. Grundy

BSc, AMICE, MBICSc
Lecturer in construction,
Salford College of Technology

Edward Arnold

8000/77/3268
690

HARTLEPOOL
COLLEGE OF
29 JUN1978
FURTHER EDUCATION
LIBRARY

© J. T. Grundy 1977

First published 1977
by Edward Arnold (Publishers) Ltd
25 Hill Street, London W1X 8LL

All Rights Reserved. No part of this publication may be reproduced, stored in
a retrieval system, or transmitted in any form or by any means, electronic,
mechanical, photocopying, recording or otherwise, without the prior permission of Edward Arnold (Publishers) Ltd.

British Library Cataloguing in Publication Data

Grundy, J T
    Construction technology.
    Vol.1
    1. Building
    I. Title
    690   TH145

ISBN 0–7131–3387–2

Text set in 10/11 pt IBM Press Roman, printed by photolithography, and
bound in Great Britain at The Pitman Press, Bath.

Contents

Preface

The design length of the TEC Construction Technology level-1 unit is 60 hours, and the content of the unit is wide and aimed at giving the student a broad appreciation of the facets of construction. In order to transfer this information to the student and to reinforce that knowledge by drawing, assignments, and other methods, I feel that some basic text is required.

This book has not been designed as a reference book, but one in which the student is given the essentials upon which the lecturer may enlarge as he feels necessary, thereby allowing more time to be spent on discussion and individual consultation rather than on laborious note-taking.

The construction industry encompasses many specialisms and the development of an interest in one's chosen field may be enhanced by an understanding of the reasons why, as well as how and when, a particular operation is performed. This interest, once developed, may result in a person becoming a life-long student.

J. T. Grundy
January 1977

Acknowledgements

The author wishes to acknowledge the assitance that he has received from his friends and colleagues during the writing of this book, and Jack Hodson for his time and effort in lettering the author's artwork.

Information and figures on wind loading, drought damage, and hardcore materials in Chapters 6 and 13 are derived from Crown Copyright material by permission of the Director, Building Research Establishment. Further Crown Copyright material in the form of extracts from the Building Regulations 1976, Advisory Leaflets, and other government publications is reproduced by permission of the Controller of Her Majesty's Stationery Office.

Extracts from British Standards are reproduced by permission of the British Standards Insitution, 2 Park Street, London W1A 2BS, from whom complete copies can be obtained.

Table 1 of the CI/SfB classifications is reproduced by permission of RIBA Publications Ltd, and figs 3.8, 3.11, 3.12, and 3.13 are reproduced from *Draughtsmanship* by permission of Fraser Reekie and Edward Arnold (Publishers) Ltd.

Information relating to electricity and gas supply is reproduced by permission of the Electricity Council and the British Gas Corporation.

The author would also like to thank TAC Construction Materials Ltd, Magnet Joinery Ltd, and Cwmbran Development Corporation for their kind assistance, and also W. H. Smith and Sons for permission to reproduce the photograph on page 123.

1   The built environment

Acknowledgement is due to the Technician Education Council for permission to use the content of the TEC units in this chapter. The council reserves the right to amend the content of its units at any time.

1.1 The elements of the built environment

The elements of the built environment consist broadly of accommodation for living, working, storage, recreation, spiritual needs, facilities for transport, and other specialist constructions. These elements in varying numbers and sizes make up the environment built by man for himself over the years.

a) *Living accommodation*   This may be divided into three groups:
   i) low-rise buildings, varying from one to three storeys in height;
   ii) medium-rise buildings, varying from four to seven storeys in height;
   iii) high-rise buildings, whose height exceeds seven storeys.

b) *Working accommodation*   This can be divided into four groups:
   i) factories,
   ii) shops,
   iii) offices,
   iv) educational establishments, from play schools to universities.

c) *Storage accommodation*   This usually takes the form of warehouses or sheds, but may also be seen as reservoirs, gasholders, and oil tanks.

d) *Recreation accommodation*   May take the form of sports centres, football and cricket grounds, swimming and athletic stadiums, or of art galleries, museums, concert halls, and cinemas. Each is fulfilling a specialist recreational role.

e) *Spiritual accommodation*   Churches and other places of worship come in this category, having provided, over the centuries, many opportunities for architects to display their skills to best advantage, Coventry and Liverpool Cathedrals being two of the most recent examples.

f) *Facilities for transport*   The provision of these has been one of the largest areas of development since the industrial revolution, the networks of roads, railways, and canals being the major elements, with shipping and harbour facilities and airport complexes being of a more localised nature.

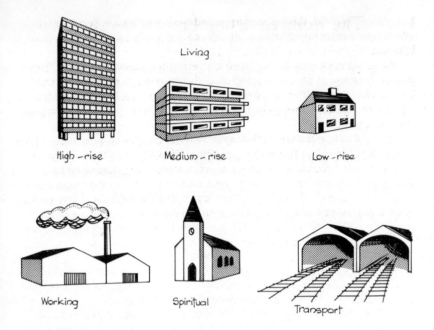

Fig. 1.1 Accommodation

1.2 Functions of the elements of the built environment

a) Living accommodation   The function of living accommodation is to provide man, his family, relations, and friends with a protective envelope against the elements. It provides a place where a person can remove himself from the remainder of society — the building providing a barrier against the 'hubbub' of day-to-day living — and also provides a base from which he can venture forth on his various activities and to which he can return at their conclusion. It is also a place of storage for his goods and chattels.

b) Working accommodation   In many of the developed and developing countries, man is dependent upon his neighbour since he is not self-sufficient in such areas as growing the food for his individual needs, making his own clothing, and manufacturing his other requirements for survival. In such countries, things are done on a far grander scale where groups of people with their own specialist talents join together to produce something which other members of the community desire. The production of many consumer items is best achieved in a stable environment, hence the need for covered working accommodation in the form of factories, mills, etc. The goods produced by the workers in a given factory will not all be required by those particular workers and the surplus can be exchanged for other goods which are required and which have been produced by other groups of workers. The more goods

2

and groups involved, the more complicated the barter system becomes, at which time a monetary system is introduced and goods are sold rather than bartered.

The goods must now be displayed in covered areas known as shops. The greater the range of goods manufactured, the more shops are required until the complex economy of today's industrial nations is reached. This system requires managing and organising, and one of the prime operations of management is communication.

Communication is achieved in one of two ways: orally or in writing. There is, therefore, a need to house the management function in offices. These are places where information can be collected, recorded, and disseminated or stored. In order that people may become useful members of the community, in whatever capacity they choose, they are required to have some knowledge of both general and specialist topics. Some of the knowledge is gained from experience, while the remainder is taught — hence the need to provide accommodation for educational purposes.

c) Storage accommodation   Living and working accommodation both involve storage, as mentioned above. Another form of storage accommodation is the warehouse, where manufactured goods may be stored until they are required by the shops, sometime in the future. For example, fireworks are manufactured throughout the year but heavy demand, far in excess of normal production, occurs at only one period of the year. It is therefore necessary to store and build up stocks during the rest of the year to meet this isolated heavy demand. Most warehouses nowadays are required to cover large areas with the minimum of internal obstruction, to facilitate modern mechanical-handling methods.

In water supply, reservoirs are used for storing water. During the wet season, excess water is stored for use during the dry season of the year.

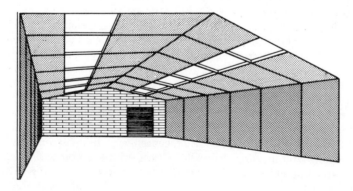

Fig. 1.2 Warehouse

d) Recreation accommodation   Man is not able to work continually for 24 hours per day, seven days a week: he needs to rest, recuperate, and change the mental and physical patterns set up at work. Recreation is one means of relieving tensions, but recreation has many facets and all must be catered for in the form of sports amenities, concert halls, stadiums, libraries, art galleries and museums, or just meeting halls. All these must be housed in some form or another.

e) Spiritual accommodation   Another way of relieving the every day nervous pressure built within a person is by spiritual means. Great efforts have been made through the ages to cater for man's spiritual needs by the erection of churches and other places of worship where he may feel at one with himself.

f) Facilities for transport   In the days before the industrial revolution, man was relatively self-sufficient and there was little need for him to travel great distances at speed; hence the means of travel was either by foot or on horseback, along dirt tracks or cobbled roadways. In a modern industrial society, people and goods travel long distances at high speed. To facilitate this, there is now a local, national, and international network of transport systems, many specialising in specific goods, e.g. oil tankers or buses. These systems in turn require special terminal buildings such as bus and railway stations, ports and air terminals, each having ancillary storage space for goods awaiting delivery or transportation.

Relationships between the elements of the built environment
The interrelationship of the various elements of the built environment in the majority of towns and cities is a result of historical development on the basis of economics and convenience. However, the recent attention to town and country planning has brought the various relationships under examination and new towns are now designed on a more selective basis, with consideration being given to the interrelationship of man's several activities.

In Great Britain, the older towns and cities are progressive enlargements of Saxon villages where the community, rather than the individual, was self-sufficient. The village was centred on the market square or green, where the bartering took place. Surrounding the green were the cottage industries and living accommodation for the farm and forestry workers. The square or green also provided the recreational area on days when no market was held, and the church was the focal point for meetings.

At the time of the industrial revolution, factories were set up either at the centres of existing population or in places where power, usually in the form of water, was readily available to drive the machinery. In the former case, the factories were built in a ring around the villages, with the more important centres attracting an influx of labour which in turn required housing, this housing being built close to the factories, thus forming another ring around the village.

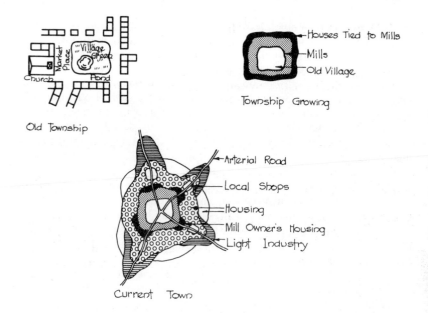

Old Township

Township Growing
- Houses Tied to Mills
- Mills
- Old Village

Current Town
- Arterial Road
- Local Shops
- Housing
- Mill Owner's Housing
- Light Industry

Fig. 1.3 Growth of Saxon village

Where factories were sited by a supply of power, in order to attract and hold a labour force the factory owners erected tied houses around the mill for their workpeople. Private enterprise produced the 'corner' shop, with all the small day-to-day needs of the isolated workforce on sale. The recreational and spiritual needs of the community were last in the order of appearance. For major items, the workman or his family still had to travel to the larger centres, where markets were held periodically.

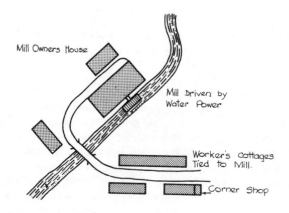

Mill Owners House

Mill Driven by Water Power

Worker's Cottages Tied to Mill.

Corner Shop

Fig. 1.4 Industrial origins

The market was, and still is, a large attraction for the surrounding population and, as the crowds were drawn to that area of a town, so the larger shops were erected in close proximity.

Since no thought was given to the effects of pollution in this period, the area surrounding the factories became very dirty. This led to the people with money erecting houses further away from the factories and on the fringe of the town, thus creating another ring. This ring, increasing in size with the population increase, was followed by the creation of local shopping centres and a further ring of light industrial accommodation which again was following the labour force, concentrating on the arterial roads.

This is the pattern of many of our present cities and towns, created as need arose and without any thought of planning. The result is that the majority of the centres, suitable for the pedestrian and the horse and cart, are unsuitable for the ever-increasing number of cars, lorries, and buses. Similarly, the railway system, designed and built during the industrial revolution for the purposes of that time, is now inadequate for modern requirements because of the constraints imposed by that earlier era.

Where 'new' towns have been erected, with advance thought and planning, the various elements of the built environment have been separated in different ways (fig. 1.5). Industrial premises are grouped together; living accommodation is grouped around small shopping centres to form neighbourhoods; the pedestrian and the motor vehicle are separated, with the roads providing more direct access from the centre to the various sectors.

1.3 Environmental considerations in building location

The many and various types of industry were traditionally centred at the most convenient position in the country; for instance, it would be thought silly to centre the fishing industry at places other than points where fishing boats could safely land their catches. Similarly, farming and farming communities are centred on the fertile river deltas and downlands; the coal industry and, following the fuel, the iron and steel industries are found in the major coalfield areas.

The climate may play a part in determining the location of an industry — such is the case with the cotton and textile industry, where damp conditions are required for spinning, thus leading to the industry being centred in the damp north-west of England.

In early days, man erected his dwellings in places where he could obtain some form of protection (fig. 1.6). He built in valleys rather than on hilltops, to obtain protection from the wind and driving rain; castles on hilltops provided protection from invaders, and settlements sprang up around the castles; at the confluence of rivers, protection was again obtained from attack.

Later, man selected his locations for convenience and profit, since they were focal points of an area. He built on the banks of a river at a point where it could be forded on foot or on horseback; at the mouths of estuaries where boats, larger than fishing vessels, could disembark their cargoes from foreign countries; at fortified camps where Roman legions had a night halt or at

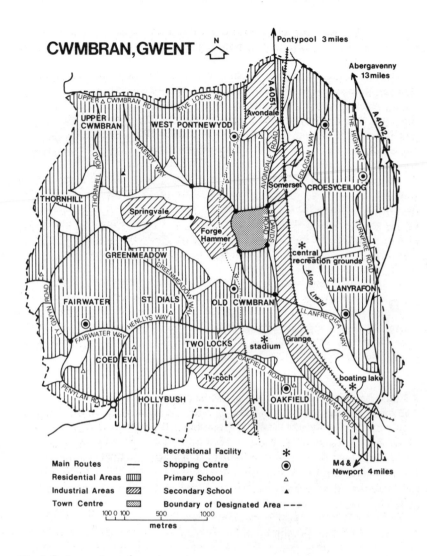

Fig. 1.5 A new-town layout

staging posts where the horses drawing the stage-coaches could be fed, watered, or changed, and travellers could have a break for food and sleep; or around market places as has been described earlier.

During the industrial revolution, man built where there was cheap power or labour readily available. Power was in the form of fast-flowing streams to drive water-wheels, timber or coal to fire boilers, or human labour for a tread-mill.

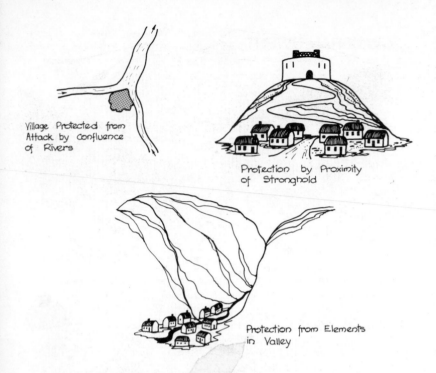

Fig. 1.6 Community positions

In modern times the siting of buildings is dependent upon
a) availability of land for building,
b) the nature of the building,
c) the proximity of services.

a) Land availability depends upon the willingness of the land-owner to sell or lease his land for building purposes. Land is usually required either in the centre of a town or city or on the periphery. In the case of the town centre, land will be available only if existing buildings are demolished. This is now being carried out in situations where the existing buildings do not comply with modern health standards or where it is economically cheaper to demolish and rebuild rather than to modify for modern industrial development.

The land around the periphery of a town is usually the most highly productive in terms of food, and with the increasing world population and demand for food it would seem a gross mismanagement of resources to take that land for building purposes rather than using less productive land elsewhere.

Early settlement

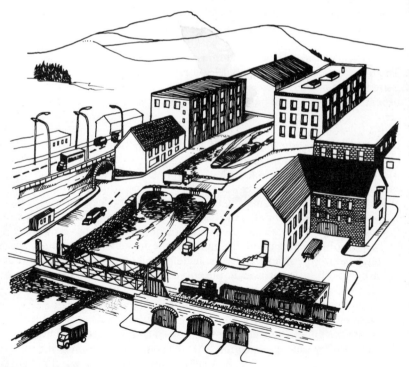

Modern town

Fig. 1.7 Development of fording place

b) The nature of the building in terms of both use and appearance is subject to strict controls by local and central government by means of green belts, planning permissions, and economic incentives.

Green belts are areas in which it is intended that only limited development of an agricultural nature may take place, thereby preserving the countryside for the benefit of the community in general.

Applications for planning consent are generally considered on the basis of

  i) does the development conform to the overall pattern predetermined for a given area?

  ii) does the proposed building blend or contrast with others in the vicinity without being an 'eyesore'?

  iii) will the building detract from the local environment or materially affect other people in the vicinity?

Grants of money may be given by central government to companies who are prepared to develop industry in certain areas, thereby assisting other industries in the area or helping to relieve unemployment problems.

c) The proximity and availability of services such as water, gas, electricity, the telephone, and sewers may determine whether a site is suitable for development, since the cost of bringing these services to the site may be economically prohibitive. All have a bearing on the suitability of a site for a given purpose.

1.4 Physical considerations in choosing a site

There are several important physical considerations which affect the choice of a site for building purposes — each will have a different degree of importance, depending on the type of building and the use to which it is to be put. They are

a) the site contours,
b) the bearing capacity and nature of the strata underneath,
c) the microclimate of the site,
d) the proximity of the site to various geographic features.

a) Man has conditioned himself to living and working on level surfaces; therefore, the more level the site, the easier and cheaper it will be to erect a building, since there will be less excavation or filling in to be carried out.

b) The size of the building and the nature of the contents will determine the weight which the ground under the building will have to support. The strata have to be strong enough to do this. In some areas the ground may move or subside as a result of geographical faults or mining operations, and costly precautions may have to be taken to overcome these problems.

c) The microclimate of a site is affected by its geographical position. Ideally, the site should be situated below the brow of a hill on a gentle southern slope, thereby gaining protection from the cold northerly winds and receiving maximum sunshine. Buildings constructed in hollows, although protected from the wind, are likely to be cold, damp, and subject to flooding — especially on low-lying ground.

d) The proximity of a site to access roads, motorways, and other transport networks; to housing or places of work; to shops; to recreational and spiritual facilities, all have a bearing on the suitability of a site for a given purpose.

2 Construction activities

Appreciates the nature of construction activities and the need for production information.
2.1 Identifies the construction site as a temporary factory employing resources of men, machines, materials and money.
2.2 Identifies drawings and associated documents used in the construction process.

Acknowledgement is due to the Technician Education Council for permission to use the content of the TEC units in this chapter. The council reserves the right to amend the content of its units at any time.

2.1 The construction site

Most people employed in manufacturing industries or commerce travel each working day to the same place of employment and, unless they change their job, that place of employment will not change for the whole of their working life. However, this is not the case for those people employed on a building site. When a particular project or section of construction or building work in which the labour force is involved is completed, there is no longer any employment for it on that site. It must then move to a fresh site which may be many kilometres away from the previous one, the distance depending on the nature and size of the building contractor employing it, or on the availability of work in a given area.

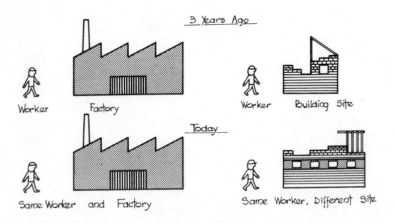

Fig. 2.1 Place of work

The construction site can therefore be regarded as a factory producing a building or buildings at a given location for a short period of time. This period does not normally exceed two years, but on large schemes it may be as long as five years.

The construction 'factory' is the same as any other factory: it converts the raw materials into a finished article by utilising resources.

The four M's

Resources may be classified under four headings:
a) manpower,
b) machines,
c) materials,
d) money.

It is the art of a good manager to utilise these resources to the best advantage.

a) *Manpower* No two construction sites are the same, and it is seldom that more than a few houses are the same. For this reason the construction operation does not lend itself to production-line methods and is therefore a labour-intensive industry.

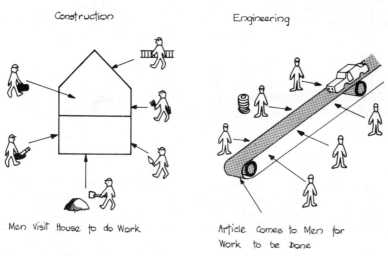

Construction

Engineering

Men Visit House to do Work

Article Comes to Men for Work to be Done

Fig. 2.2 Methods of production

The labour force is comprised of people possessing a vast range of skills, from concretor to plumber, joiner to asphalter, each in his turn going to the site and performing his specialist operation, rather than the work moving to the employee as on a production line.

Forecasting the demand for labour on a site, and its organisation once there, can create many problems.

b) *Machines*   As the pace of life has increased, so has the demand for the fast erection of buildings. This can be achieved by the use of machines wherever conditions will allow. As with the labour force, there is a wide range of machines, from concrete-mixer to circular saw, excavator to crane, and there is a range of capabilities in each area, so the selection of the right machine for the job is of prime importance if optimum speed of erection is to be achieved.

c) *Materials*   The majority of materials used in construction are natural ones. Their correct selection and use is of the utmost importance for the appearance and durability of the building. Raw materials are worked and bonded together to form the structure, while other materials come on to the site in their finished form, such as sink and cupboard units, ready for inclusion in the structure.

d) *Money*   The person who is having a new building constructed is investing a large amount of money in the project and he is entitled to expect value for his outlay in the form of durability balanced with a minimum of maintenance. The builder requires money to purchase the materials for the construction, to pay the wages of the labour force, and to hire or purchase the machines and equipment for the proper and speedy erection of the work.

The more efficiently the builder manages the resources, the more will be his profit on the cost of undertaking the work. If this were not the case, the builder would be better off investing his money resources in a bank or other similar institution for a fixed return or interest.

2.2 Construction organisation and documentation

When individuals or a company, known as the *employer*, require a new building, they seldom have the ability to design and construct it themselves. They therefore employ specialists in these fields.

The specialist designer is the *architect*, a person who will assess his client's requirements and the possibilities of a chosen site.

At this stage the architect will seek *outline planning permission*. This is the submission of an application form and site plan to the local authority in order to obtain officially the planning authorities' reaction to his client's proposals prior to the preparation of detailed drawings or even the purchase of a site. If the proposed development is an industrial or office building, it is often necessary to obtain an *Industrial Development Certificate* or *Office Development Permit* from the Department of Trade and Industry. These will be issued only if the Department considers that the local regional or national balanced distribution of industry will not be affected. The architect will also check with the various *statutory undertakings* that the various services — water, electricity, etc. — will be available to the site.

The architect next prepares sketch drawings of how he visualises the building should appear both outside and inside, for approval by his client or the building owner and from which a *quantity surveyor* can prepare an approximate cost estimate.

Having obtained the client's approval of the proposals, the architect now seeks *detailed planning permission* as required under the Town and County Planning Acts. This relates to the siting and appearance of the building and to its design. Once approval is given, he can proceed with the preparation of further drawings.

In order that the building can be erected, the architect has to communicate his intentions to the *builder*. This is done in a number of ways.

a) Working drawings   These are detailed plans, elevations, and sections, drawn to suitable scales, which indicate the position and form of the whole building and its components, together with information about the proposed site. These drawings can be classified as follows.

Location drawings
 i) Block plans — to identify the site in relation to the locality.
 ii) Site plans — to position the proposed buildings on site, together with information on proposed road, drainage, and services layouts, and other site information such as levels and strata details.
 iii) Location plans — to position the various areas within the building and to locate the principal elements and components.

Component drawings
 i) Ranges — to show the basic sizes and reference system of standard components.
 ii) Details — to show the information necessary for the manufacture of components.

Assembly drawings — to show in detail the junction between elements, between elements and components, and between components.

b) Specification (fig. 2.3)   This is a precise description of the standard of materials and workmanship which the architect requires for the work. For small building works it may be incorporated on the drawings, in the form of a description, but for a large contract it is usually a separate document. Many building materials and components are subject to *British Standards*, and there are several *British Standard Codes of Practice* which cover workmanship. The British Standards Institution, the publisher of these documents, is a private body receiving financial assistance from central government and representing all interested parties on a given topic. The standards laid down are the minimum which are acceptable and are the result of much research and deliberation in technical committees. Reference to these standards reduces the work involved in writing a specification.

Another standard which may be used is that of an *Agrément Certificate* issued by the Agrément Board. This board is an independent body which assesses new building products, developing its own methods of assessment and testing, and providing prospective users of a new product with performance information not previously available from an independent source.

15

CEMENT
MORTAR

30. Cement mortar shall consist of Portland cement and sand
as previously specified in the following proportions :

For brickwork and kerbs: 1 volume of Portland cement to
                                   3 volumes of sand.

For rendering and pipe joints: 1 volume of Portland cement to
                                   2 volumes of sand.

Sand for the above mortars is to comply with Table 2 of BS 1200.

The ingredients shall be measured in proper boxes and
thoroughly mixed on a clean watertight platform three times in
a dry state and again while sufficient water is being added
through a rose for a sufficient number of times to produce a
mixture of suitable consistency or otherwise they shall be
mixed in an approved mixer. No mortar which has developed its
initial set or has become dirty shall be used but shall be
rejected and condemned and no softening or re-tempering of
mortar which has set or hardened will be permitted.

WATER

31. Only fresh clean water free from organic or mineral
impurities drawn from the water mains or other sources approved
by the Architect shall be used for mixing cement, grout, mortar
or concrete.

HARDCORE

32. Hardcore used shall consist of broken stone or other
approved hard material clean and free from extraneous matter.

CLINKER
ASH

33. Clinker used shall be approved hard well-burnt furnace
clinker containing a good proportion of large pieces and free
from waste metal, dust, china or other rubbish.

Fig. 2.3 A typical specification

c) Schedules   These give tabulated information on a range of similar items
which may differ in detail, e.g. windows.

d) Bill of quantities (fig. 2.4)   This document gives a complete description
and measure of the quantities of labour, materials, and other items required to
carry out the work. It is prepared, usually by a quantity surveyor, from the
drawings, specification, and schedules, and gives each contractor offering to
do the work the same information on which to base his *tender*. There is the
added advantage that individual items are priced in the tenders.

The way in which the items are described and measured is layed down in
the *Standard Method of Measurement for Building Works* (issued by the Royal
Institution of Chartered Surveyors and the National Federation of Building

Item No.	Description	Quantity	Unit	Rate	Amount
			SUPERSTRUCTURE		
	BRICKWORK AND BLOCKWORK				
	BRICKWORK				
	Common bricks as described in gauged mortar (1:1:6)				
A	Half-brick wall	149	m^2		
B	Half-brick wall in skin of hollow wall	292	m^2		
C	Reduced brickwork in attached piers	18	m^2		
D	Form cavity 50 mm wide between skins of hollow wall including wall ties 5 per m^2	281	m^2		
E	Close cavity horizontal at sill by returning outer skin of facing bricks as brick on edge and including vertical damp-proof course 112 mm wide	8	m		
F	Close cavity vertically at jambs by returning outer leaf of facing bricks and including vertical damp-proof course 112 mm wide.	22	m		

Fig. 2.4 Extract from a bill of quantities

Trades Employers) and the *Standard Method of Measurement for Civil Engineering Quantities* (issued by the Institution of Civil Engineers and the Federation of Civil Engineering Contractors).

Before work can begin on the site, the architect must obtain approval from the relevant local authority that the proposed works conform to the requirements laid down by central government.

The main requirements are contained in the *Building Regulations 1976* (in London, the London Building (Constructional) By-laws). These regulations involve the completion of application forms describing the proposed work and submission of drawings showing block plans, plans of every floor and roof, and sections of every storey to a scale of not less than 1:100; structural details; and details of services and fittings.

Other specialist requirements are covered by:
a) for factories — the *Factories Act*, the *Clean Air Act*, the *Thermal Insulation (Industrial Buildings) Act;*
b) for offices and shops — the *Offices, Shops and Railway Premises Act;*
c) for hotels and catering establishments — the *Food Hygiene Regulations*.

There are other statutes which cover fire-fighting appliances and means of escape from fire in buildings.

Having obtained the necessary approvals under these regulations, the architect will send the drawings, specification, and bill of quantities, together with the *Conditions of Contract*, to contractors for them to submit tenders. The quantity surveyor will inspect the tenders returned and will advise the client, through the architect, of the most suitable contractor. The client and the selected contractor then sign the contract, in which the contractor agrees to

17

carry out the construction work and the client agrees to pay a sum of money for that work.

When work has begun on the site, the contractor must give written notification to the local authority at various stages of the work. This allows an official of the authority to inspect the work, thereby ensuring that it is being carried out in accordance with the approved drawings and regulations. On large jobs the architect may employ a *clerk of works* who is permanently on the site to look after the client's interests.

During the course of the construction, the client may decide that he wants certain alterations made to the original scheme, or the architect may find that one of his proposals requires modification. In both cases an *architect's instruction* or *variation order* is issued to the contractor in writing, sometimes with amended drawings, amending the original contract — the cost of the variation being agreed between the quantity surveyor and the contractor.

As has been shown earlier in this chapter, a large amount of money is required by the contractor to finance a construction operation. To reduce the capital outlay, especially on large jobs, the quantity surveyor will make a periodic assessment of the work done by the contractor and of the quantity of materials purchased by him but not yet incorporated into the building. On the basis of these interim valuations the architect will issue *Interim Certificates*, authorising the client to make payments to the contractor.

After completion of the contract, for which the architect must issue a *Certificate of Practical Completion*, the quantity surveyor and the contractor agree upon the final account. However, this account is not paid in full because a certain amount of money, known as *Retention*, is held back for an agreed period. This period, usually six months, is to allow for defects resulting from work not being in accordance with the specification to become apparent, and is known as the defects liability period. At the end of this time, the contractor, having rectified any defects to the architect's satisfaction, is entitled to receive the retention monies.

3   Drawing

Appreciates the importance of drawings and sketches as a means of communicating technical information.

3.1 Prepares freehand sketches of constructional details to approximate scale, in orthographic and isometric projections with and without guide lines.
3.2 Uses scales to obtain correct proportions for freehand sketching.
3.3 Uses BS 1192 and BS 308 as sources of standard symbols and notation.
3.4 Interprets information from production drawings using the preferred scales.
3.5 Identifies the main types of drawing used in the construction process.
3.6 Prepares dimensioned drawings using instruments.

Acknowledgement is due to the Technician Education Council for permission to use the content of the TEC units in this chapter. The council reserves the right to amend the content of its units at any time.

It can be seen from the previous chapter that the drawing is an essential means of communicating technical information in the construction industry, whether the drawing is formal or in sketch or pictorial form. It is therefore essential that the student should develop this means of communication to an acceptable standard — as has often been said, a drawing or sketch can be worth a thousand words.

3.1 Types of projection and lettering

There are many methods of drawing a particular object, each method showing some different facet of that object. The student will eventually have to decide for himself which method to use when presenting a particular piece of information. The problem which has to be overcome is that of showing an item which is three-dimensional on a piece of paper which is only two-dimensional.

This can be done by means of *projection* — the drawing of a given object as seen from a particular viewpoint or a number of viewpoints. There are several methods of projection, but the two which are most commonly used in the construction industry are both known as *orthographic projection* and are called (a) *first-angle* and (b) *isometric projection*.

The student will appreciate that the drawing of many constructional details to the same size as the finished item would be impractical. It is therefore necessary to reduce, or *scale* down, the finished size of the item for drawing purposes, but it is essential that the various dimensions be reduced by a similar amount so that the appearance is not distorted. In the initial stages of sketching, it is convenient to use guide lines or graph paper so that proportionality can be maintained. Once the student has mastered this technique, the guide lines may be dispensed with.

a) *First-angle projection*

This method of showing constructional details involves looking at a given object from a number of viewpoints, all mutually perpendicular, and drawing what can be seen from a particular position.

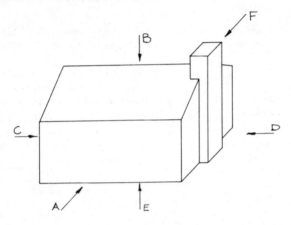

Fig. 3.1 Outline of a house

The outline of a house shown in fig. 3.1 can be conveniently viewed in the direction of the arrows in order to show all the relevant information. However, the view in the direction of arrow E is seldom used. The view from A is called the front elevation; from C and D the side elevations; from F the rear elevation; and from B the plan.

These views are set out in a particular way (fig. 3.2). The front elevation (from A) is drawn first and the other views are related to it. The view from above (B) is placed underneath; the view from the left (C) is placed on the right; the view from the right (D) is placed on the left; the view (F) from the rear may be placed on the further left or right, depending upon convenience; the view from below (E) is placed above.

b) *Isometric projection*

This method is used for presenting the detail in pictorial form, and the principle in this projection is that all vertical lines remain vertical, while all horizontal lines are drawn at an angle of 30° to the horizontal (fig. 3.3). Any inclined lines are difficult to draw, but by the use of guide lines, and by treating that section as initially part of a solid, the detailing is made easier. It

20

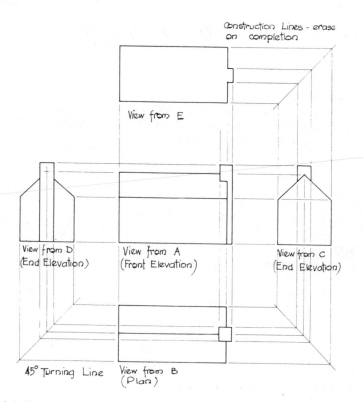

View from E

View from D
(End Elevation)

View from A
(Front Elevation)

View from C
(End Elevation)

45° Turning Line

View from B
(Plan)

Fig. 3.2 First-angle orthographic projection

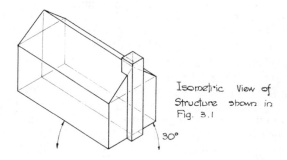

Isometric View of Structure shown in Fig. 3.1

30°

Fig. 3.3 Isometric projection

should be noted that none of the sloping side lengths will be in the same proportions as the remainder of the detail.

This projection is frequently used to show working processes or 'exploded' views of components (fig. 3.4).

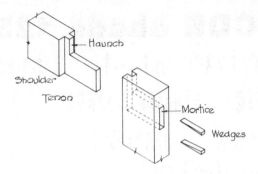

Fig. 3.4 Exploded view of haunched mortice-and-tenon joint

Lettering

A drawing without any description has little meaning; therefore, if it is to be a useful communication tool, there should be some annotation. The standard of lettering on a sketch can greatly detract from or improve its communicative value, depending upon legibility and position. The aim should be to make neat, easily readable notes which do not obscure the drawn details.

There are numerous styles of lettering (fig. 3.5) and it is only with practice that a good standard can be achieved. The student will eventually adopt his own style, but it is suggested that a simple style is initially adopted and practiced using guide lines.

3.2 Scales

If true proportionality is to be achieved in sketching and drawing, it is essential that all dimensions be reduced in the same ratio, i.e. one unit of measurement on the drawing represents another unit of measurement in practice. The relationship of these two measurements, expressed as a fraction, is known as the representative fraction or *scale*. For example, if a line 1 millimetre long represents a dimension of 1 metre, the scale would be $\frac{1}{1000}$ or 1:1000, since 1 mm represents 1000 mm (= 1 m).

In both sketching and drawing it is essential that the correct scale is used, and there are three limiting factors when determining this:
a) the size of the drawing sheet, bearing in mind the desirability of keeping the sheets for one project to one size, or the space available on the sheet for inclusion of the detail;
b) the size of the detail in relation to the amount of information to be imparted;
c) the need for economy of time and effort in the preparation of the detail.

3.3 Drawing conventions

It has previously been stated that a drawing is used to transfer information from designer to builder, but too many drawings may tend to confuse and the

ABCDE abcde 12345

ABCDE abcde 12345

ABCDEFG   abcdefg 12345

ABCDE   12345

abcdefg

𝔄𝔅ℭ𝔇𝔈 abcde 12345

Types of Transfer Lettering

A B C D E F   a b c d e   1 2 3 4 5

A B C D E F   a b c d e   1 2 3 4 5

A B C D E F   a b c d e   1 2 3 4 5

A B C D E F   a b c d e   1 2 3 4 5

Types of Freehand Lettering

Fig. 3.5 Lettering

number of drawings may be reduced if one drawing can do the work of two. This can be achieved by using some form of 'shorthand' and, to ensure that one person can understand another person's drawing, British Standards BS 1192, 'Building Drawing Practice', and BS 308, 'Engineering Drawing Practice', give recommendations for the standard presentation of information.

a) *Lines* (fig. 3.6)

Type of line		*Purpose*
————————	Thick	Site outline, new building, primary elements, profile of component
————————	Medium	Existing buildings, general details, secondary elements, outlines of components
————————	Thin	Reference grids, dimension lines, hatching
– – – – – – – –	Medium dotted	Hidden details
—·—·—·—	Thick chain ⎫	
—·—·—·—	Medium chain ⎬	Pipelines
—·—·—·—	Thin chain	Centre lines

Fig. 3.6 Lines

b) *Dimensions*

If a building is to be properly constructed, dimensions should be clearly shown on the drawings, together with the points to which they relate. The type of arrowhead on a dimension line indicates a particular type of dimension (fig. 3.7):

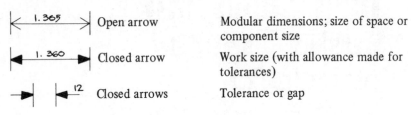

1.365	Open arrow	Modular dimensions; size of space or component size
1.360	Closed arrow	Work size (with allowance made for tolerances)
12	Closed arrows	Tolerance or gap

Fig. 3.7 Dimensioning

The written dimensions should be on top of the dimension line, preferably in a central position and expressed in terms of millimetres or metres. Where the two units of measurement are to be used on the same drawing, dimensions in metres should be expressed to three decimal places (e.g. 5.6 m should be shown as 5.600 m). Normally, whole numbers indicate a dimension in millimetres, thus avoiding the symbol mm.

c) *Symbols*

A symbol is a form of abbreviation. BS 1192 contains a large number of standard symbols which represent

i) service components, fixtures, and fittings, together with features on the ground (fig. 3.8);
ii) types of material (fig. 3.9), especially when that material has been cut to show a section — this type of symbol is referred to as *hatching*;
iii) words or phrases which would otherwise result in lengthy notes on the drawing (fig. 3.10).

It is only by practice in reading drawings and recognising the standard 'shorthand' that the student will be able to extract the information he requires.

3.4 Preferred scales

There are several types of working drawing (referred to in the previous chapter) and BS 1192 gives preferred scales for each type.

a) *Location drawings*

i)	Block plans	1:2500
		1:1250
ii)	Site plans	1:500
	(fig. 3.11)	1:200
iii)	Location plans	1:200
	(fig. 3.12)	1:100
		1:50

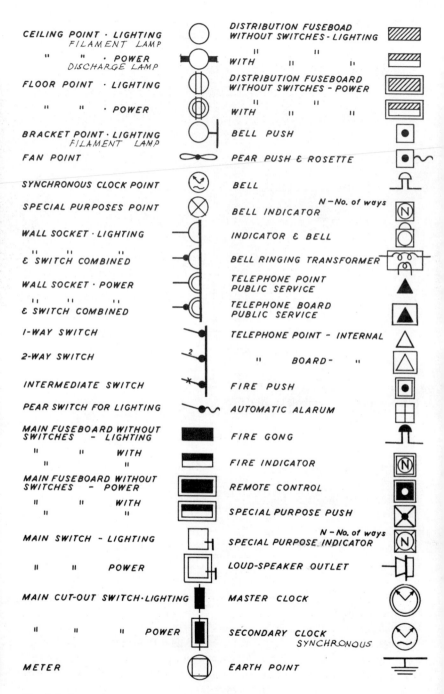

Fig. 3.8 Drawing symbols (components) (see also BS 3939 and PD 6479)

Material	Symbol	Material	Symbol
Brick		Metal	I L
Concrete		Partition Block	
Earth		Plywood	
Fibre-board		Screed	
Glass		Sheet Membrane	
Hardcore		Stone	
Loose Insulation		Sawn Wood / Wrot Wood	

Fig. 3.9 Drawing symbols (materials)

Aggregate	agg	Hardcore	hc
Bitumen	bit	Inspection chamber	IC
Boarding	bdg	Invert	inv
Brickwork	bwk	Mild steel	MS
Building	bldg	Plasterboard	pbd
Cast iron	CI	Polyvinyl chloride	PVC
Column	col	Rain-water head	RWH
Concrete	conc	Rain-water pipe	RWP
Cupboard	cpd	Reinforced concrete	RC
Damp-proof course	DPC	Rodding eye	RE
Damp-proof membrane	DPM	Softwood	swd
Drawing	dwg	Tongue and groove	T&G
Foundation	fdn	Vent pipe	VP
Granolithic	grano	Wrought iron	WI

Fig. 3.10 Standard abbreviations

26

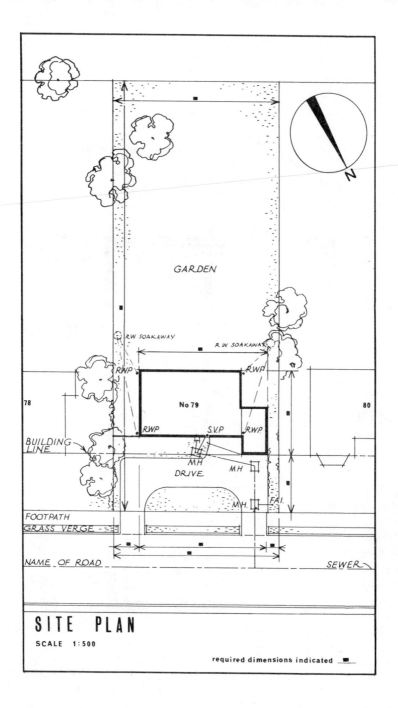

Fig. 3.11 Site plan

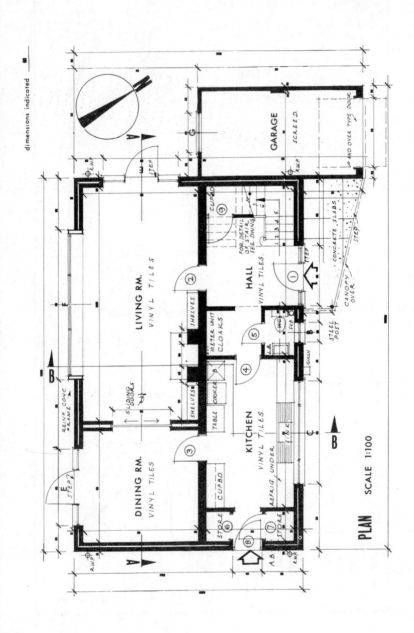

Fig. 3.12 Location plan

28

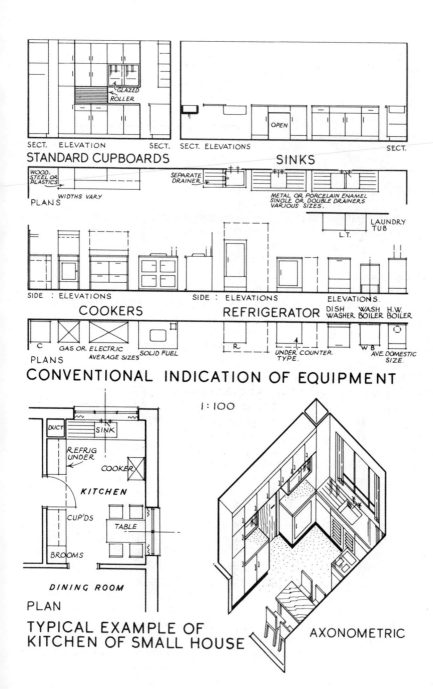

STANDARD CUPBOARDS

SECT. ELEVATION   SECT.

GLAZED ROLLER

SINKS

SECT. ELEVATIONS   SECT.

OPEN

WOOD, STEEL OR PLASTICS

PLANS   WIDTHS VARY

SEPARATE DRAINER

METAL OR PORCELAIN ENAMEL
SINGLE OR DOUBLE DRAINERS
VARIOUS SIZES.

LAUNDRY TUB

L.T.

COOKERS   REFRIGERATOR

SIDE : ELEVATIONS   SIDE : ELEVATIONS   ELEVATIONS.

DISH WASHER   WASH BOILER   H.W. BOILER

C   GAS OR ELECTRIC
          AVERAGE SIZES   SOLID FUEL

R   UNDER COUNTER TYPE.

W B   AVE. DOMESTIC SIZE.

PLANS

CONVENTIONAL INDICATION OF EQUIPMENT

1:100

DUCT   SINK
REFRIG UNDER   COOKER
KITCHEN
CUP'DS   TABLE
BROOMS

DINING ROOM

PLAN

TYPICAL EXAMPLE OF
KITCHEN OF SMALL HOUSE   AXONOMETRIC

Fig. 3.13 Ranges

29

b) *Component drawings*

iv)	Ranges	1:100
	(fig. 3.13)	1:50
		1:20
v)	Details	1:10
		1:5
		1:1

c) *Assembly drawings*

vi)	Assembly	1:20
		1:10
		1:5

3.5 Preparing drawings

The preparation of a drawing rather than a sketch implies the correct use of equipment, materials, and procedures.

a) *Equipment* (fig. 3.14)

i) *Ruling pens* These consist of a handle with two steel blades of equal length, held together by an adjusting screw. Ink is inserted between the blades, the pen being held vertically, and the line thickness is adjusted by the screw.

ii) *Drawing ruling pens* Similar to a fountain pen, and having interchangeable heads or nibs for various line thickness.

iii) *Compasses* Used to draw large-diameter circles, they comprise two self-centring arms approximately 150 mm long attached to a handle at one end. At the free end of one arm there is a divider point, while at the end of the other arm there is the facility for taking a pencil, a ruling pen, or divider points.

iv) *Spring-bow compasses* These are used to draw small-diameter circles and they are smaller than compasses (75 mm arms). They have a sprung ring at the handle end of the arms, and the arms, acting on a pivot, are held together by an adjustment screw.

v) *Set squares* Used to draw vertical and inclined lines, and made of transparent plastics, they are triangular in shape and have square or bevelled edges. There are three types of set squares, the first having angles of 60°, 30°, and 90°; the second having angles of 45°, 45°, and 90°; the third having one angle of 90° and the facility to adjust the other two angles.

vi) *French curves* Used to draw curves of changing radius, they are made of transparent plastics and have either square or bevelled edges.

vii) *Drawing boards* Used to provide a flat surface on which to draw, they are made in varying qualities and sizes: the quality varying from faced blockboard with plastics edges to spruce boards with aluminium edging; the size of the boards being suitable to accommodate standard paper sizes (A2 – 420 mm x 549 mm; Al – 549 mm x 841 mm; AO – 841 mm x 1189 mm).

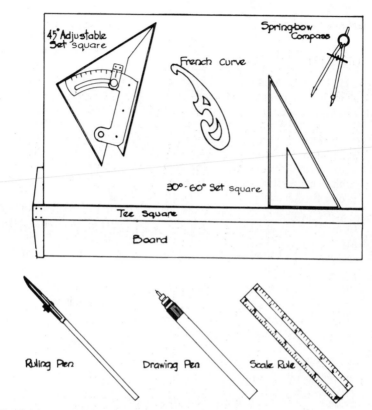

Fig. 3.14 Drawing instruments

viii) *Tee squares* Used to draw parallel lines and made of plastics or mahogany with plastics working edges.

ix) *Stencils* Used to provide guides for lettering and made out of transparent or tinted plastics, they provide a range of lettering styles and sizes. Special pens are usually required for use with the stencils.

x) *Erasing shields* Used to erase one particular line from a complicated drawing without affecting other lines in close proximity, they are made from thin steel.

xi) *Scale rules* Used to plot dimensions or to ascertain dimensions from drawings, they are made of plastics or wood in 150 or 300 mm lengths. The number and scope of the scales varies depending on the shape of rule and the type of drawing generally prepared.

The student may find other specialist items of equipment useful, though not essential.

b) *Materials*
   i) *Drawing paper*   There are many qualities of this paper in the form of
      cartridge paper (governed by BS 1343) produced in rolls or cut to
      standard sizes. A finer paper, known as detail paper (governed by BS
      1342), may be used for preliminary sketches and drawings and is suit-
      able for certain reprographic processes.
   ii) *Tracing paper or film*   As with drawing paper, there are many
      qualities, the requirements being contained in BS 1340. Surface finish
      can be smooth or matt, with the latter wearing a pencil point away
      much more quickly. Plastics film is the more durable of the two
      materials and should be used for record purposes. When applying ink
      to the surface, it is advisable to remove surface grease by dusting with
      talc or French chalk, to allow an even flow of ink.
   iii) *Pencils*   Pencil leads are produced in a variety of hardnesses, ranging
      from 9H to 5B, and the most suitable are H or 2H for detail lines, HB
      for writing, and B for lining in and sketching. The ends should be
      sharpened to a chisel point for drawing lines and to a round point for
      writing or sketching.
   iv) *Ink*   Drawing ink can be obtained in a range of colours and is con-
      tained in bottles or tubes which should be kept closed unless in use.
      The ink is waterproof and can, therefore, be used with colour washes.
      Certain colours of ink have poor reprographic qualities.
   v) *Draughting tape*   This should be used for holding paper in position
      on the drawing board, since pins will damage the board surface and
      Sellotape will remove the surface layer of paper. Clips may be used
      where paper and board have the same nominal size.
   vi) *Erasers*   A soft eraser or art gum should be used for the removal of
      pencil work, while a hard eraser or a razor blade will be required for
      ink work.

c) *Procedures*
   i)   Ensure that hands and instruments are clean before starting work —
        dirty hands and equipment will leave a dirty drawing.
   ii) Set paper on board, lining up with a tee square, and draw margins
        and a title box. In industry these are frequently preprinted on the
        drawing or tracing sheet in a standard format.
   iii) Plan the positioning of the elevations, plans, and details to be drawn
        on the sheet. This planning operation may determine either the scale
        to which these views can be drawn or the number of details which
        may be drawn without loss of clarity. The location of the details
        should follow some logical sequence.
   iv) Draw the outlines of a particular item before drawing the precise
        detail, using a 2H or H pencil.
   v)   Line in the outlines using a B pencil.
   vi) Print the attendant notes.
   vii) Erase all unnecessary guide lines.

viii)   Erase other mistakes. When ink work has been erased from an area, it is advisable to rub the surface with a smooth hard object such as a pen end, cow's tooth, or thumb nail before redrawing over that area; this will prevent ink lines 'blotching'.

The student should always bear in mind that the drawing is an essential means of communication within the construction industry, and each line drawn should have something to say to the person 'reading' the drawing.

4   The evolution of the structure

Identifies the constitutent parts of a structure and describes their evolution.

4.1 Identifies substructure, superstructure and primary elements.
4.2 Identifies secondary elements and finishings.
4.3 Distinguishes between self finishes and applied finishes.

Acknowledgement is due to the Technician Education Council for permission to use the content of the TEC units in this chapter. The council reserves the right to amend the content of its units at any time.

There are two main parts to any structure — the *substructure* and the *super-structure* (fig. 4.1). The substructure is usually considered as that part which is below ground level or just above ground, such as the damp-proof course, and the superstructure is that part of the construction which is above the substructure.

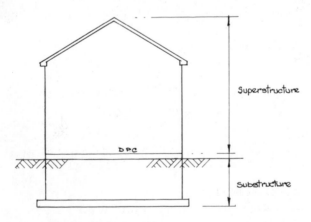

Fig. 4.1 Substructure and superstructure

The superstructure, and to a lesser extent the substructure, comprises various constructional elements (fig. 4.2) which are classified into primary, secondary, and finishing elements.

The SfB system (fig. 4.3) classifies the various elements and gives each a reference number, e.g. a handrail is a secondary element with the reference

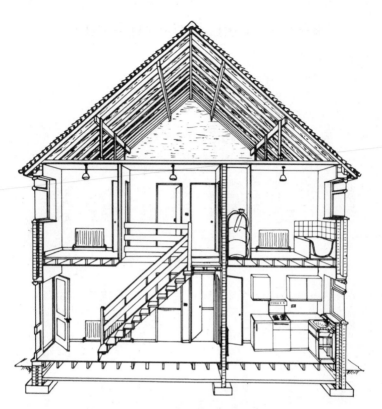

Fig. 4.2 The elements of a building

number 34. The system also classifies the constructional operations by upper-case letters, e.g. F for brickwork, and the material content by lower-case letters and numerals, e.g. n5 for rubber. The SfB system, widely used in the construction industry for referencing government, technical, and trade information, derives its title from the Samarbetskommitten för Byggnadsfrågor, the Swedish name for the committee which devised the system.

4.1 Evolution of the building structure

The substructure of a building comprises the foundations, walls, and floors which provide the support for the remainder of the building.

In early days there was no need for substructures since buildings had no foundations, and the need for foundations came about only when the weight of a building was such that the sides started to sink into the ground. This weight increase was, in the main, due to the increase in height of buildings and the inclusion of more than one floor. Foundations, in early times, consisted of taking the wall of the building down to a level below ground which was stronger than that at the surface. The idea of spreading the load was then

Substructure	Superstructure			Services		Fittings		Site
(1–) Ground, substructure	(2–) Primary elements	(3–) Secondary elements	(4–) Finishes	(5–) Mainly piped	(6–) Mainly electrical	(7–) Fixed	(8–) Loose	(9–) External elements
(10)	(20)	(30)	(40)	(50)	(60)	(70)	(80)	(90) External works
(11) Ground	(21) External walls	(31) External openings	(41) External	(51)	(61) Electrical supply	(71) Circulation	(81) Circulation	(91)
(12)	(22) Internal walls	(32) Internal openings	(42) Internal	(52) Drainage, waste	(62) Power	(72) Rest, work	(82) Rest, work	(92)
(13) Floorbeds	(23) Floors	(33) Floor openings	(43) Floor	(53) Liquid supply	(63) Lighting	(73) Culinary	(83) Culinary	(93)
(14)	(24) Stairs, ramps	(34) Balustrades	(44) Stair	(54) Gases supply	(64) Communications	(74) Sanitary	(84) Sanitary	(94)
(15)	(25)	(35) Suspended ceilings	(45) Ceiling	(55) Space cooling	(65)	(75) Cleaning	(85) Cleaning	(95)
(16) Foundations	(26)	(36)	(46)	(56) Space heating	(66) Transport	(76) Storage, screening	(86) Storage, screening	(96)
(17) Piles	(27) Roofs	(37) Roof openings	(47) Roof	(57) Ventilation	(67)	(77) Special activity	(87) Special activity	(97)
(18)	(28) Frames	(38)	(48)	(58)	(68) Security, control	(78)	(88)	(98)

Fig. 4.3 The SfB system

36

introduced and consequently large flat stone slabs were positioned in the bottom of the wall trench. As bricks took over from stone as the main structural building material, they were initially founded on the stone slabs, but later the brick wall was widened at its base by the use of footings. Eventually concrete strips replaced the footings as the artificial foundations of walls (fig. 4.4).

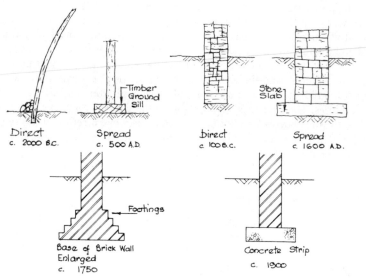

Fig. 4.4 Development of foundations

The floors of a building were originally just compacted earth, but in times of heavy rain this became muddy and uncomfortable, so stone slabs were used to overcome this problem. The stone, however, was cold to walk on, and the Romans, by designing a method of under-floor heating (fig. 4.5), created the suspended floor. At a similar period in history, the need for more living

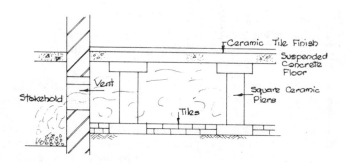

Fig. 4.5 Roman under-floor heating (the hypocaust)

accommodation on the same floor area created the suspended timber first floor, timber being used because of its strength, relatively light weight, and ready availability.

The external appearance and development of the primary elements such as walls and roofs, together with the secondary elements that are incorporated in the walls, such as doors and windows, have been linked with the development of architecture from Saxon and Norman times to the present day. The development of the elements is most clearly mirrored in church construction, since it was the churches, having the finances and resources available, which used architecture in its fullest sense — design with a view to aesthetic appeal. Architecture until more recent times has, however, been dependent upon the local raw materials and the tools available for use on the construction, while domestic buildings mirror the social, industrial, and economic changes through the centuries. The change is most notable in the growing up of the hut, the decline of the castle and manor house, and an eventual meeting of the two types of construction to form a composite at approximately the end of the sixteenth century.

English architectural periods

1800 BC	Prehistoric, stone age
	Bronze age
	Iron age
43–420 AD	Roman
420–1066 AD	Saxon ⎫
1066–1189 AD	Norman ⎬ Romanesque
1189–1307 AD	Early English or English Gothic
1307–1377 AD	Feudal or Decorated
1377–1485 AD	English or Perpendicular
1485–1558 AD	Tudor ⎫
1558–1603 AD	Elizabethan ⎬ early renaissance
1603–1625 AD	Jacobean ⎭
1625–1702 AD	Stuart ⎫
1702–1810 AD	Georgian ⎬ late renaissance Greek revival
1810–1820 AD	Regency ⎭
1820–1875 AD	Victorian ⎫ Industrial
1875–1920 AD	Post 1875 Housing Act ⎬ revolution
1920–date	Modern

It should be noted that the periods stated are approximate, since changes take place over a period of time and no one architect or monarch can govern such changes.

Early forms of house were 'A-frame' arrangements which were covered by animal skins; later the frame was formed by two forked uprights supporting a horizontal ridge pole with closely spaced rafters having their feet bedded into the ground and their tops lashed to the ridge (fig. 4.6). The covering was of interwoven branches, plastered with mud and covered with straw, heather, or

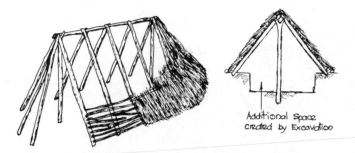

Fig. 4.6 A primitive dwelling

Additional Space created by Excavation

reeds; headroom was increased by digging the enclosed space to a depth of some 0.5 m, and holes were left in the ridge to allow smoke to escape.

Anglo-Saxon building (fig. 4.7) required larger and better houses but the same basic framework remained. The increased size meant increased weight of covering, which led to large sizes of timber being required and additional ground loading. As a result, the rectangular timber framework appeared, in which the joints were made by squaring off the timber members, and the facility of a chimney was also incorporated at one side. Saxon towers and halls were built of stone, and their windows and doorways originally had triangular heads, later semicircular, with straight splayed sides and short cylindrical columns.

The Norman period (fig. 4.8) evolved the cruck frame for the small hall, with the pairs of halved trees having a natural curve supporting the ridge and connected at a suitable height by a tie beam. The tie beam was later carried past the cruck to provide support for a wall-plate which in turn supported the feet of the rafters; thus the rafters were not taken down to ground level and the walls, the area between wall-plate and ground sill, were filled with stonework, or wattle and daub set on a close vertical timber framework. The Norman church had a massive central square stone tower; the windows remained narrow with semicircular heads, but with ornate carvings. To carry the heavy roofs and stone floors, vaulting was introduced, as was the collar roof, where pairs of rafters were tied together.

A feature of Gothic architecture (fig. 4.9) was the pointed arch over door and window openings. Further height and weight of both walls and roofs to churches brought about the introduction of the buttress, which counteracted the thrust of the arch, while the columns became more ornate. The larger spans required of timber roofs resulted in support for the tie or collar beam in the form of a brace, horizontal purlins at collar level, and wind bracing between purlin and cruck or main rafter. It was during this time that the English squire started to have his bedroom at first-floor level, and the cottage also started to take the now conventional form of a rectangular plan, having single-storey stone-rubble walls faced and topped with a sand, clay, and straw mixture. Openings for doors and windows were timber posts and lintel, and the

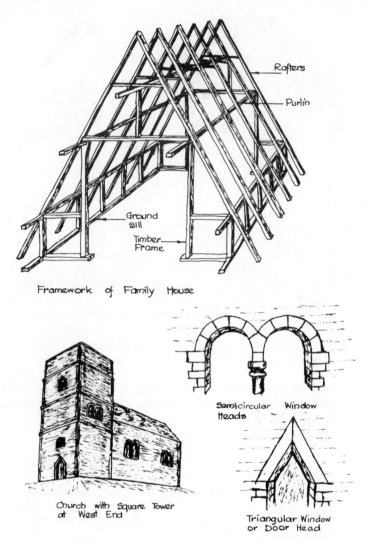

Rafters

Purlin

Ground
Sill

Timber
Frame

Framework of Family House

Semicircular Window
Heads

Church with Square Tower
at West End

Triangular Window
or Door Head

Fig. 4.7 Saxon building

roof was of thatch on a light timber frame construction, with the fireplace
and chimney placed at one end. The later part of the early-English period saw
more ornate stone arches with multiple curves.

Moving on to the fifteenth century (fig. 4.10), the peasants' house remained
of cruck construction, but the springing up of a new middle class with new-
found status increased the development of the two-storey timber-frame house
construction. The focal point was still the main central hall which rose to the
full height of the house and was flanked at either end by the two-storey con-

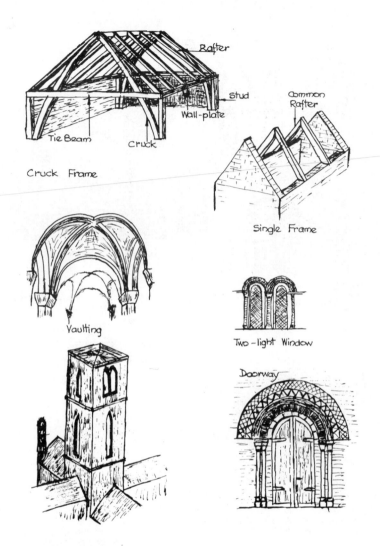

Fig. 4.8 Norman building

struction. Stone gables were becoming common and provided stability to the structure. The roof framework remained visible from the inside and, while strength was still required, beauty also started to play a part, with braces becoming more ornate and a more complex system of lighter-weight members forming the frame. Square and rectangular windows in relatively long lengths now started to appear, with churches having spires and their windows widened to provide increased light inside a wider church. These windows, while still having pointed arches, started to contain and then develop tracery, and orna-

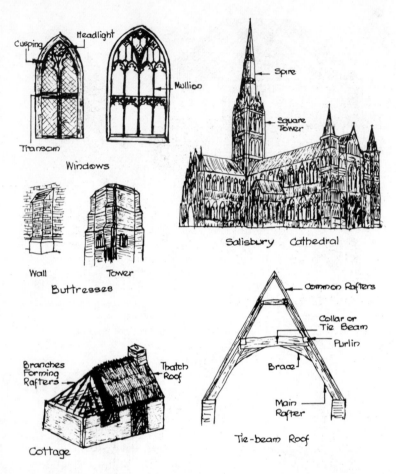

Fig. 4.9 Gothic building

mental ribbing was added to vaults. In this period, wood-carving was starting to come into its own, with motifs similar to those of the stonework.

Larger churches led to yet larger windows with the arch still prevalent, but the larger windows were split up into smaller units with transom bars included. The arches became flatter towards the end of the fifteenth century and fan vaulting started to appear, with roof pitches becoming flatter or steeper — the former being masked by a parapet while the latter were of hammer-beam construction requiring flying buttresses to support the thrust of the roof on the external walls.

In the sixteenth century (fig. 4.11) the planning of the house became important, with emphasis on privacy and pleasant well-lit rooms. Although

Studs.
Bearer
Joists

First-floor
Overhang

Bedroom
Kitchen   Main Hall   Bedroom
Store

Collar
Queen Post

King
Post

Arched - Braced   Queen Post   King Post
Roof   Framework

Flying
Buttress   Hammer-beam
Roof

Vaulting

Flatter Arch Window

Fig. 4.10 Fifteenth-century building

windows were still comparatively small, the timber mullions and transoms began to appear. Chimneys became a feature rather than an afterthought, being built higher and in ornamental fashion, such as the twisted-brick variety. The roof space became part of the living area, resulting in the appearance of dormer windows in walls, and the timber framework of closely spaced studs became more open with further horizontal members and diagonal bracing.

Ordsall Hall with fifteenth-century oriel window

The churches of the period took on a perpendicular rather than a squat style, with architects incorporating features of the old Greek and Roman architecture in the decorative work.

The Elizabethan period saw bricks, previously a rare and expensive building commodity, coming into more general use for the houses of the middle and upper classes, and the timber framework was filled in with bricks, frequently in herring-bone pattern, which were then covered with plaster. The windows became flat-headed with diamond-shaped small panes joined with strips of lead and fixed into an iron framework. It was during this period that the domestic staircase made its appearance, rather than a step-ladder access to the loft or first floor. Roof coverings of red clay pantiles and stone became more common than the thatch roof, because of their durability.

During the seventeenth century (fig. 4.12), brick and stone replaced timber as the load-bearing materials of domestic constructions, and, as the strength of the materials was realised, so the height of buildings grew to three and four storeys. Windows were enlarged and towards the end of the century were almost storey height. Great architects such as Inigo Jones and Sir Christopher Wren produced a multitude of fine churches and stately homes, and straight lines and proportion became the elevational vogue. The bonding of brickwork developed and the floor beams — laid on edge rather than flat, for greater strength — were built into it. Chimneys, which for two centuries had remained on the external walls, were positioned to suit the fireplace, each fireplace having a separate chimney. Internally, plaster was frequently applied to the walls or, as an alternative, wooden panelling would be used to mask damp

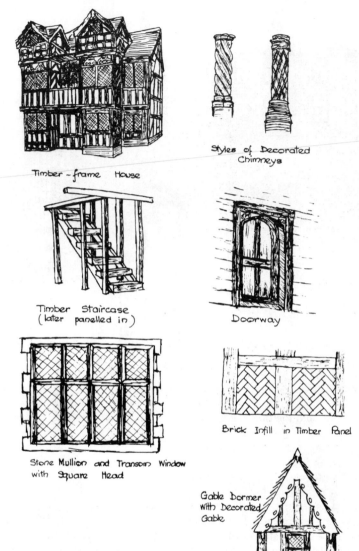

Timber - frame House

Styles of Decorated Chimneys

Timber Staircase (later panelled in)

Doorway

Stone Mullion and Transom Window with Square Head

Brick Infill in Timber Panel

Gable Dormer with Decorated Gable

Fig. 4.11 Sixteenth-century building

patches. Resulting from the sealing up of the crevices in walls by plaster, ventilation was reduced and the opening window was developed. Originally horizontally sliding, which suited the earlier seventeenth-century design, it was later developed into a vertically sliding sash in a box frame, which was more in keeping with the later vertical appearance. Doorways were given special treatment as a feature, having stone framework and elaborate heads. In the

45

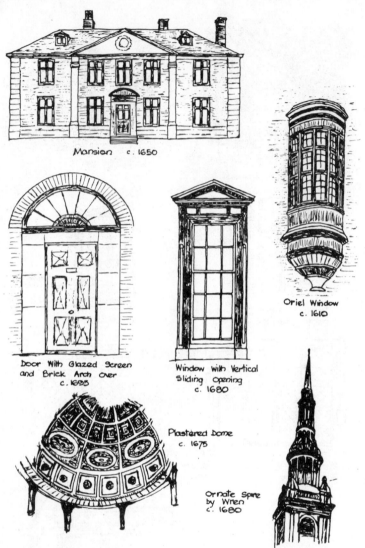

Mansion c. 1650

Door With Glazed Screen
and Brick Arch over
c. 1695

Window with Vertical
Sliding Opening
c. 1680

Oriel Window
c. 1610

Plastered Dome
c. 1675

Ornate Spire
by Wren
c. 1680

Fig. 4.12 Seventeenth-century building

south-eastern regions of England, the hanging of clay tiles on a timber-framed upper-storey construction provided good weather resistance.

The eighteenth century (fig. 4.13) saw the entrance-door features developed even further, with glazed screens being incorporated above the door itself, but the vertical appearance of the facade was still prevalent, and chimneys had again moved back to the outside walls, usually the gable. The arch again

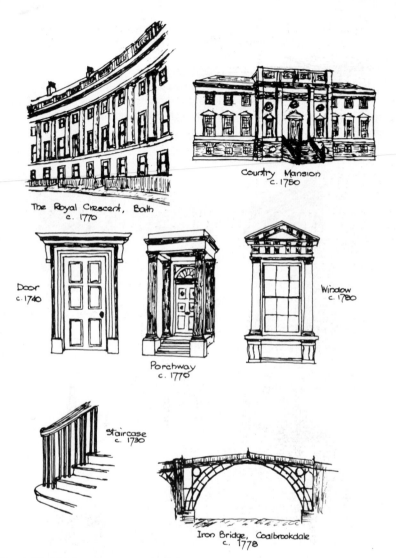

Fig. 4.13 Eighteenth-century building

appeared at the window head, but this time formed of brickwork. During this period there were three distinct types of building erected: (a) the country mansion built on a vast rectangular scale with colonnade facades, (b) long rows of town houses, three- and four-storeys high and including basement scullery quarters and servants' quarters in the loft, and (c) workers cottages, situated close to mills and factories and built 'back-to-back' in long rows, of two- and later three-storey construction having a single room at each level.

47

The pitched roof began to change to a hipped roof or, on smaller properties, to the mansard roof (see fig. 10.1(b)(iv)) which gave increased headroom on the top floor.

The Regency period (fig. 4.14) continued the development of the entrance with porticos and porches which were matched in the elevational treatment by bow and bay windows. The internal finishes, especially the plastering which gave an ornate ceiling finish and the wooden panelling, were works of art in themselves.

Town Houses

Cottages

Aqueduct Near Llangollen c. 1800

Fig. 4.14 Regency

The industrial revolution (fig. 4.15) played an important part in the development of architecture by making available new materials and mechanising the production of the older ones such as bricks. Slates became available throughout the country as a result of the transport revolution and were the most widely used roof covering. Their use affected the slope of the roof, and a pitch of 30° was found to be satisfactory. The walls remained solid, and chimneys were constructed to serve two or more fireplaces. With the manufacture of large panes of glass, the small-pane windows went out of fashion, but the basic vertically sliding sash method of opening remained. The houses of the workers remained in long terraces but, as the population increased, and with it the demand for housing, five-, six-, and seven-storey tenement blocks were erected close to city centres. It was not until this period that washing and toilet facilities were incorporated into the house, and it was the Public

48

Manchester Town Hall c. 1870

Flats c. 1880.

Terraced Housing c. 1840

Kings Cross Railway Station c. 1850

Mill c. 1885

Clifton Suspension Bridge c. 1850

Fig. 4.15 Nineteenth-century building

Health Act of 1875 which first laid down minimum standards not only for hygiene, in the form of water supply, drainage, and sewage disposal, but also for wall thickness, damp-proof courses, and street widths and lengths. These regulations produced the typical rows of houses which can still be seen in many towns and developed as a result of the industrial revolution. Another architectural feature of the age was the development of the curved structure seen in concert halls and in railway architecture, especially station roof and bridge structures.

The twentieth century could be classed as the age of experiment, in which architects experimented by combining the development of (a) materials, (b) structural-design concepts, and (c) industrialised production methods in varying proportions, with space and shape (fig. 4.16). This rapid development

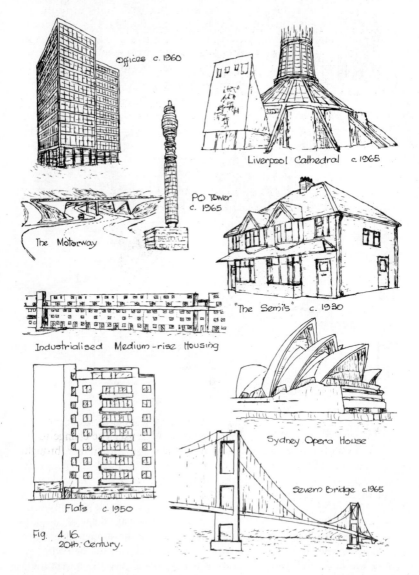

Fig. 4.16 Twentieth-century building

has been facilitated by the slow painstaking development of earlier centuries and the recent rapidly changing political and economic pressure bringing about changes of emphasis in technology.

4.2 Evolution of secondary elements and finishes
The secondary elements of a structure are those which are incorporated in a building once the primary elements of the structure have been completed, and these elements, with the exception of door ironmongery, were not really developed until the sixteenth and seventeenth centuries, but then kept pace with the development of the superstructure. The gables of buildings received various treatments, but from Tudor and Elizabethan times the roof outline was emphasised by large, sometimes ornate, barge- or weather-boards.

Rain-water disposal from a roof did not create any problems until the end of the eighteenth century, when terraced houses were built directly alongside a thoroughfare and caused inconvenience to the pedestrian. The water was initially collected behind a parapet and was discharged by means of a spitter or gargoyle at one end of the building (fig. 4.17) – the gutter as we know it today came with the development of materials during the industrial revolution.

Fig. 4.17 Gargoyle

The development of finishings in the true sense again dates from the sixteenth century, when timber panelling was in wide use. The change to plaster as a wall finish, partially as a result of the scarcity of timber, brought about a mixture of the two materials, with the timber framework being left exposed and the plaster providing the infilling on a suitable backing. This eventually led to the areas of high risk, as far as damage was concerned, continuing to be made in timber, i.e. skirtings and door and window surrounds, while the remainder of the area was plastered to a smooth surface. During the ornamental Georgian and Regency periods, the plasterer became almost a sculptor, creating designs on both walls and ceilings, but more functional considerations of the later nineteenth and twentieth centuries relegated this finish back to a plain surface.

Floor finishes were stone, wood, or carpet until the late nineteenth century, when newer materials were developed or became more readily available. The discovery of plastics has probably effected the biggest change in floor finishes, not only in the finish material but also in the laying and fixing of other more established floor and wall finishes.

4.3 Types of finish

An *applied finish* is one which is deliberately applied to a material component or unit, either during the manufacturing process, prior to inclusion in the building, or on completion of the fixing operations. Examples of this type of finish are

a) paints applied to any surface,
b) sand-facing or glazing applied to a brick surface,
c) plastics laminate applied to the top of a kitchen cupboard unit as a working surface,
d) rendering to a wall.

A *self finish* is one which is a direct result of the manufacturing process. Examples of this type of finish are

a) the smooth surface and colour achieved when producing plastics materials,
b) the smooth surface resulting from the planing and sanding of wood,
c) the surface finish of a paint film once applied, e.g. matt or gloss.

HARTLEPOOL
COLLEGE OF
29 JUN1978
URTHER EDUCATION
LIBRARY

5 Excavation work

Understands the need for excavation work on construction sites.

5.1 Identifies substructure.
5.2 Explains reasons for the removal of vegetable soil.
5.3 Identifies and describes various types of excavation.
5.4 Explains need to remove water from excavation.

Acknowledgement is due to the Technician Education Council for permission to use the content of the TEC units in this chapter. The council reserves the right to amend the content of its units at any time.

5.1 The substructure

In the previous chapter the substructure has been identified as that part of the building which is below ground level and supports the superstructure (fig. 5.1). While occasionally providing storage facilities in small buildings, and being utilised for car parking and service facilities for such items as boilers and air-conditioning plant in larger buildings, the main purpose of the substructure is to transfer the weight of the superstructure to the surrounding ground.

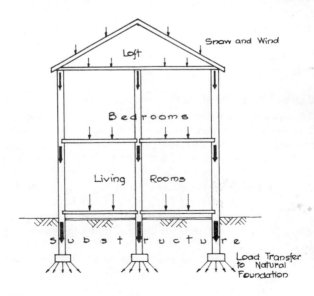

Fig. 5.1 The substructure

In order to construct the substructure, it is first necessary to excavate the ground until some stratum is reached which will support the total weight of the building when it is fully occupied.

5.2 Removal of topsoil

Unless the work is being carried out in a redevelopment area, the first section of ground to be excavated is the topsoil or vegetable soil. The Building Regulations (clause C2) state that 'the site of any building, other than an excepted building, shall be effectively cleared of turf and other vegetable matter'. The reason for this is that the top layer of ground contains material which is decomposing, along with active plant life in the form of grass, flowers, roots or shrubs, etc. This material is easily compressible and as such is unsuitable for foundations.

The topsoil layer is approximately 150 mm to 300 mm deep and has been built up by nature over thousands of years; it cannot be replaced overnight and is therefore a valuable commodity. This layer is excavated separately from the other excavation work and the material is stockpiled in some convenient position on site in order to provide gardens on completion of the building work. Alternatively, the soil is sold to people who have only a thin layer of topsoil on their land and who wish to increase its depth in order to improve the productivity of their gardens.

5.3 Types of excavation

There are several types of excavation used in the building process, the type and method depending on the amount of ground to be excavated. Broad divisions are
a) soil strip,
b) reduced-level excavation,
c) bulk excavation,
d) trench excavation,
e) hole or pit excavation.

a) Soil strip   It has already been stated that the vegetable soil must be removed before any other excavation takes place. The process is to strip the surface of the ground of its topsoil covering, and this may be achieved by the use of certain types of machinery or plant (fig. 5.2):
  i) by a *bulldozer* pushing the topsoil layer to a convenient stockpile position a short distance away;
 ii) by a *scraper* 'paring off' the top layer, transporting it, and depositing it at a stockpile position some distance away on a large site;
iii) by a *drag line* scraping off the top layer and either depositing it at a convenient stockpile or loading the material into waggons for transport off site;
 iv) by a *backactor* operating in a similar manner to the drag line;
  v) by a *mechanical shovel* (track- or wheel-driven) operating in a similar manner to the drag line;
 vi) by a *skimmer* – a machine specifically designed for this type of work.

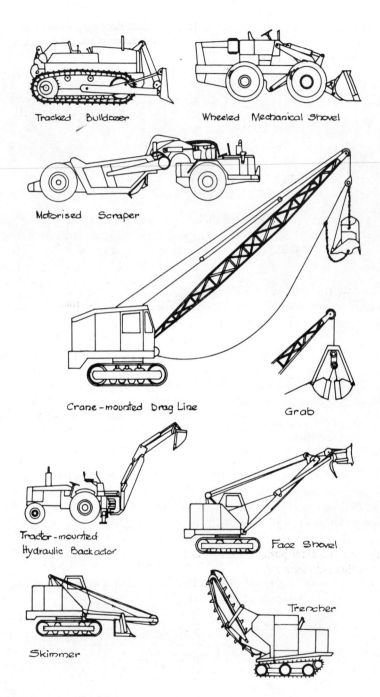

Tracked Bulldozer

Wheeled Mechanical Shovel

Motorised Scraper

Crane-mounted Drag Line

Grab

Tractor-mounted Hydraulic Backactor

Face Shovel

Skimmer

Trencher

Fig. 5.2 Excavating machinery

Semimobile hydraulic backactor excavator

b) Reduced-level excavation (fig. 5.3)   The ground which lies between the topsoil and the Earth's crust is known as the subsoil and comprises particles of weathered rock of various shapes and sizes. It is usually necessary to provide a level surface from which construction may take place, and this level may be lower than the top of the subsoil. In such cases, excavation into the subsoil is required to reduce the level of the ground. The volume of subsoil to be excavated, together with the nature of the material, will determine the type of machine and methods to be used.

On a sloping site it may be necessary to cut into the banking on one side of the proposed building, while filling with suitable material (possibly the material 'cut' out) on the other side, in order to achieve the level working surface or formation level. This method of excavating in one area and transport-

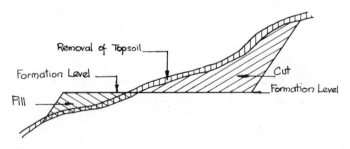

Fig. 5.3 Reduced-level excavation

ing and using the same material for filling purposes is known as 'cut and fill' and is frequently used in the construction of new roads.

The machinery used is similar to that outlined in section 5.3(a).

c) Bulk excavation   Where large volumes of subsoil are required to be excavated in order to reach the formation level, the excavation is known as bulk excavation. This type of excavation may be to reduce levels or to provide basement areas. In addition to the plant already mentioned, a face shovel may also be used for this type of work.

d) Trench excavation (fig. 5.4)   Having excavated or filled to a formation level, it is frequently necessary to excavate trenches to a lower level. In these trenches the foundations of the building are constructed. Trenches are also required outside the building for the laying of pipes and cables which carry gas, water, electricity, and other services.

Where the trenches are straight and uniformly shaped, it is generally cheaper to use a machine for the excavation — trench hoes, wheel ditchers, and back-actors are the ideal machines to use in such situations. However, where the trenches are small and have a complicated layout, a gang of men with picks and spades can still provide the most economical method of excavation.

If very deep trenches are required, i.e. lower than the reach of a backactor, a grab or clamshell mounted on the jib of a crane may be used.

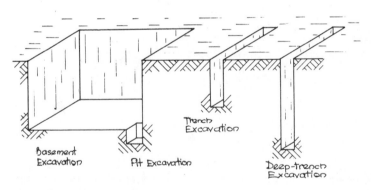

Fig. 5.4 Types of excavation

e) Hole or pit excavation   The design of the substructure may require holes or pits to be excavated. In the case of circular holes, hand or mechanically driven augers or drills are used. Pits, being small shallow or rectangular holes, may be excavated mechanically or by hand.

5.4 Water in excavations
There is always water present in the ground, and the level at which it rests is known as the standing water table (fig. 5.5).

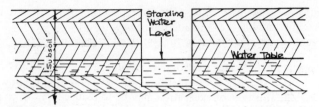

Fig. 5.5 Water table

If excavation is to be carried out to a depth below that of the level of the water table, the excavated area will be filled with water. This is not desirable, since

a) it is difficult to carry out further construction work in water without incurring extra expense, especially when workmen rather than machines are involved;

b) the ground on either side of the excavated area is less stable and more likely to collapse when waterlogged (fig. 5.6).

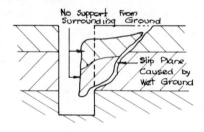

Fig. 5.6 Unstable ground

If the excavation depth does not reach the level of the water table, there may still be a problem of water in the bottom of the excavation, especially in heavy rain, because the excavation may act as an open drain (fig. 5.7). This provides an easier route for any rain-water falling on the surrounding ground to reach the water table, rather than the natural percolation route which is followed in the absence of excavation work. Any workman walking on the bottom of a wet excavation will disturb the ground with the imprint of his boot, and this disturbed ground will have to be removed before any work in the excavation takes place, thereby increasing the depth of the excavation, the volume of filling material, and hence the cost of the operation.

It is therefore essential that any water which enters the excavation should be removed as soon as possible — or a better alternative is to prevent the water entering in the first place.

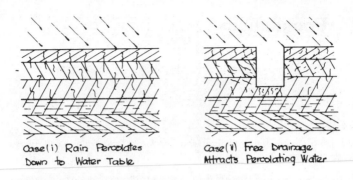

Case(i) Rain Percolates Down to Water Table

Case(ii) Free Drainage Attracts Percolating Water

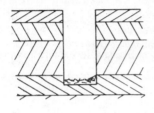

Loose Wet Ground in Trench Bottom to be Cleaned off

Fig. 5.7 Trenches affected by water

6 Foundations

Understands the functions of a foundation, factors affecting choice and fundamentals of construction.

6.1 States primary functions of a foundation and identifies associated terminology.
6.2 Explains primary factors involved in the choice of a foundation.
6.3 Explains how subsoil movements can affect a foundation and describes how these may be minimised.
6.4 Identifies the primary materials used for foundation construction and explains their suitability.
6.5 Identifies pad, slab and strip foundations and describes their construction.
6.6 Identifies foundations to beds and pavements and describes their construction.

Acknowledgement is due to the Technician Education Council for permission to use the content of the TEC units in this chapter. The council reserves the right to amend the content of its units at any time.

6.1 Functions of a foundation

The Building Regulations, clause D3(a), require all foundations of buildings to

'safely sustain and transmit to the ground the combined dead load, imposed load and wind load in such a manner as not to cause any settlement or other movement which would impair the stability of the building or any adjoining buildings'.

The dead load is the self weight of the building, which includes the substructure, superstructure, primary and secondary elements, and finishes. The imposed load is the weight of the occupants, furniture, and movable goods, together with the weight of snow which may rest on the roof. The wind load is the force which the wind, blowing from any direction at an assessed maximum velocity, will exert on the building. (See fig. 6.1.)

A foundation has two components: the natural foundation and the artificial foundation (fig. 6.2). The natural foundation is that part of the subsoil on which the structure rests, and the artificial foundation is that part of the structure which transmits the load to the natural foundation.

There are many types of subsoil, each having some ability to carry load. This ability is known as the bearing capacity and is defined as the safe load which a unit area of ground will carry. The bearing pressure is the amount of force that a structure exerts on the ground per unit area; hence the bearing

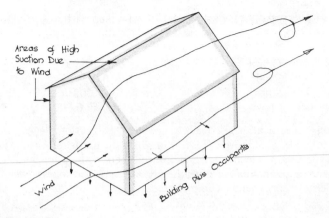

Fig. 6.1 Loadings

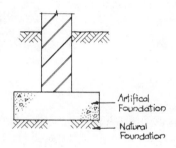

Fig. 6.2 Natural and artificial foundations

pressure must not exceed the bearing capacity of the ground if the structure is to remain standing.

Subsoil type	Bearing capacity (kN/m²)
Rocks (granite to chalk)	10 000–600
Compact gravels and sands	600–300
Stiff clays	400–200
Soft clays	100–50
Loose gravels and sands	200–75
Very soft clays and silts	75–0
Peat and made ground	To be determined by tests

Fig. 6.3 Table of typical bearing capacities

The various subsoil types can be classified by their particle or grain size.

Subsoil type	Particle size (mm)
Gravel	More than 2.0
Sand	2.0–0.06
Silt	0.06–0.002
Clay	Less than 0.002

This size-classification also leads to another classification, namely cohesive or non-cohesive. A cohesive soil is one in which the particles have the ability to stick together, while in a non-cohesive soil there is no such adhesion. The gravels and sands are non-cohesive; the silts and clays are cohesive.

Gravels and sands can be compressed only slightly from their natural state, and the settlement of these materials under a load occurs in a very short period of time. The particles have voids between them, but the volume of voids in comparison to the total volume of the material is generally small (fig. 6.4). However, these voids will allow water to flow through the material with ease, but the presence of water will not cause any great change in the volume of the material as a whole.

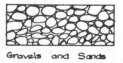

Gravels and Sands   Silts

Fig. 6.4 Particle and void sizes

Silts and clays are highly compressible, but settlement under load takes a long time. The particles again have voids between them, but, although these are small, the proportion of voids in the total volume is large because the particle size is small. Since the voids are small, water does not easily flow through the material, and because the particles are smaller there is considerable change in volume when water is present.

6.2 The choice of a foundation

The choice of a foundation depends on

a) the ability of the artificial foundation to support the weight of the structure together with all the imposed loads and to transmit those forces to the natural foundation;

b) the ability of the natural foundation to carry those loads without undue deformation — the major factors in determining the ability of the natural foundation to carry a load are the type of subsoil strata under the building and the amount of water present in those subsoils;

c) the ability of the structure to sustain small movements at foundation level without detriment to its overall load-carrying capacity — some buildings are designed to be stiff and rigid, while others are designed to accommodate small movements at foundation level.

6.3 Ground movement

Clause D3(b) of the Building Regulations states that foundations shall be taken down to such a depth, or be so constructed, as to safeguard the building against damage by ground movement. Ground movement may result from a number of factors.

a) *Subsidence due to mining underground* (fig. 6.5)   As tunnels are driven underground, support to the ground above is provided by props, rings, or columns of rock. The props may be removed on completion of the working, or the tunnels may be filled with waste products, thus in time the props will corrode or rot, the columns will collapse, and the strata above the tunnel will settle. If the building is situated where subsidence is likely, it must be designed to allow for such movement.

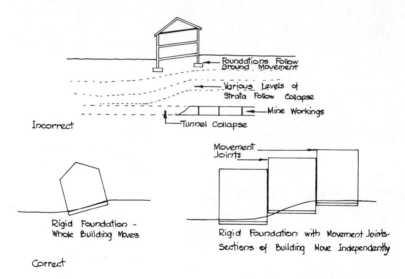

Fig. 6.5 Subsidence

b) *Overloading of the natural foundation* (fig. 6.6)   This will cause the subsoil in the region of the foundation, or of a section of the foundation, to be deformed. This in turn will cause a section of the artificial foundation to move, and with it a section of the building superstructure.

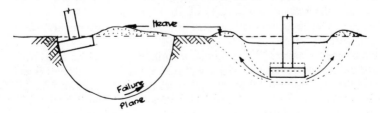

Fig. 6.6 Overloading

c) *Lowering of the ground-water table*   When water is removed from the subsoil, the voids which the water previously occupied will become empty. Subsoils having a large grain size will not be affected, but those having a small grain size, i.e. a high proportion of voids, will settle as the grains endeavour to fill up the empty voids.

The water table may be lowered in a number of ways:
 i) by pumping from wells or boreholes for water supply,
 ii) by the insertion of field or land drainage,
 iii) by excavating below the level of the water table,
 iv) by roots drawing water to feed trees and plants (fig. 6.7),
 v) by lack of rainfall causing river and stream levels to fall.

In the case of (i), (ii), and (iii) it is the responsibility of the persons carrying out the work to ensure that their actions do not have a detrimental effect on surrounding property, and in case (iv) it is advisable to avoid having trees closer to a building than their mature height, especially poplars,

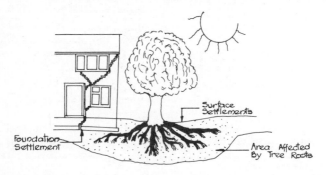

Fig. 6.7 Removal of ground water by tree roots

whose root systems may affect the ground for a distance of 25 m around them. In case (v) the only answer is to construct the foundation at a level such that the seasonal variation of the water-table level does not fall below the level of the foundation.

d) *Increase in the water-table level*   This may cause certain soil types to lose some of their cohesion and therefore their load-bearing capacity. Alternatively, it may be desirable to reduce the water-table level in order to increase the load-bearing capacity of a soil type. In either case the water table may be reduced locally by inserting land drains around or under the structure.

e) *Freezing of moisture in the ground*   During frosty periods, moisture in the ground above the water table will freeze. Water expands as it freezes and this expansion will force the soil particles apart, causing a condition known as frost heave. In certain circumstances this condition may lift the foundation of a building. In severe winters in Great Britain, the frost may penetrate the ground to a depth of 600 mm, therefore the foundations of a structure should either be taken down to a level at which the subsoil is not affected by frost heave or be designed to withstand that effect.

f) *Settlement*   If the weight of the building which rests upon an area of the natural foundation is more than the weight of earth which previously rested on that area, a certain amount of settlement can be expected and accepted, providing that a similar settlement occurs over the whole foundation area. Differential settlement will occur if there is change in the nature of the natural foundation strata and this must be taken into account in the design of the foundation. This may be accomplished by either spreading the load over a wider area or by strengthening the foundation so that it will span any 'soft spots'.

6.4 Foundation materials

The artificial foundation, i.e. that part of the substructure directly in contact with the natural foundation, is being highly compressed by the loads from the structure above and by the reaction of the ground beneath; also, the soil which surrounds the artificial foundation may contain water and other chemicals. The material which is used in the construction of the artificial foundation must therefore be strong in compression and must not deteriorate as a result of chemical attack or the presence of water.

In olden days, locally quarried rock or stone blocks were used for foundations, but in certain areas where the rock is weak (chalk, limestone, sandstone) this material would not support heavy loads.

Bricks will fulfil the requirements of a foundation material and, with the improvement in quality of the brick in the nineteenth century, many brick foundations were constructed. The spread of the brickwork at its base in order to spread the load of the wall to the natural foundation is known as a 'footing' (fig. 6.8).

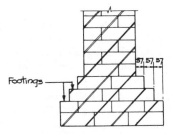

Fig. 6.8 Footings

There are, however, two drawbacks to brick foundations. Firstly, there is the problem of excavating the bottom of a trench to provide a level surface on which to lay the bricks, a level surface being essential to ensure that the courses of brickwork are horizontal. Secondly, brickwork is only as strong as the mortar which acts as the binding agent between the bricks, and, although

Portland cement was 'discovered' in 1824, its use with sand to produce mortar did not supersede the previously used lime mortars until the beginning of the twentieth century.

The increase in the size and weight of modern structures required an increase in the strength of the foundation materials. Concrete, a material made possible by the discovery of Portland cement, was the material which filled that requirement, being very strong in compression although weak in tension. The four basic ingredients of concrete are cement, fine aggregate, coarse aggregate, and water, which when mixed together set to form a hard stone-like material, the strength depending on the proportions of the materials in the mix. Concrete poured into the bottom of a trench can be spread and finished to give a level upper surface, suitable for the laying of bricks (fig. 6.9).

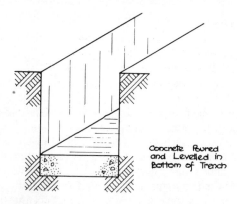

Concrete Poured and Levelled in Bottom of Trench

Fig. 6.9 Strip foundation

The material will resist the attack of most chemicals with the exception of sulphates. Where these occur in the ground, their attack can be nullified by the use of special cements such as 'sulphate-resisting' or 'high alumina'. Once concrete has set, water will not affect its strength, but the presence of too much water during the mixing and setting periods will seriously reduce the strength of the material.

Steel was used for certain types of foundation in the early part of the twentieth century, but because of its cost and the problems of corrosion it was superseded by concrete.

6.5 Types of foundation
The most common types of foundation in use today are strip, slab, and pad foundations.

a) Strip foundations (fig. 6.10)   The majority of low-rise buildings are constructed in such a way that the external walls transmit the loads of the floors and roof to the foundations. This load is fairly uniformly distributed along

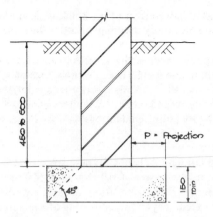

Fig. 6.10 Strip-foundation detail

the length of the wall, so the foundation to the wall is in the form of a strip of concrete having uniform width and thickness.

The Building Regulations (table to clause D7) give suitable minimum widths of strip foundations for various loadings and types of subsoil. The concrete strip should project an equal amount on either side of the wall so that there is an even distribution of the load on to the subsoil.

The concrete thickness must not be less than its projection from the base of the wall and in no case less than 150 mm. This is because concrete tends to crack under excessive loads at an angle of $45°$, but if the projection is equal to the thickness there will be no reduction of the area of concrete bearing on the subsoil.

A modern variation of the strip foundation is that of the deep-strip or trench-fill foundation (fig. 6.11). Where it is necessary to excavate to a depth

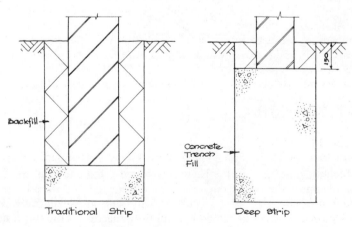

Fig. 6.11 Deep strip foundation

67

of 900 mm or more, in order to found the building on suitable strata, or to avoid excessive ground movement, it may be possible to reduce the width of the foundation. This in turn will reduce the amount of soil which must be excavated. By filling the trench with concrete to a level some 150 mm below ground level, more concrete will be used, but the costs of brick, mortar, and labour involved in laying the bricks are greatly reduced, thus producing an overall saving of both time and money. A similar saving can also be made if the depth of an ordinary strip foundation is increased to a level 150 mm below ground level.

b) Slab or raft foundations (fig. 6.12)   These are used where the load-bearing capacity of the subsoil is very weak (soft ground or fill) or where there is a likelihood of subsidence. These foundations cover the whole area of the building, and sometimes extend beyond it, consisting of a concrete slab, strengthened by steel reinforcement, up to 400 mm thick, with extra thickening under load-bearing walls. The slab is constructed near ground level, unless a basement is incorporated in the building design, and, in order to protect the ground near the perimeter of the raft from the effects of the weather, a deeper edge beam is constructed or a concrete paving is laid around the building.

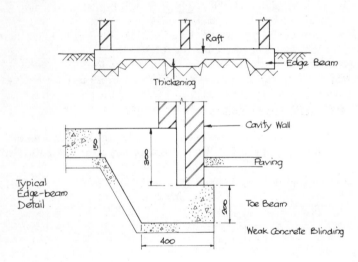

Fig. 6.12 Raft foundation

c) Pad foundations (fig. 6.13)   These are isolated slab foundations which support columns. The most economical form of construction is to provide each column with its own base, the column being set centrally on that base. The thickness of the slab should be equal to the projection of the slab from the side of the column.

68

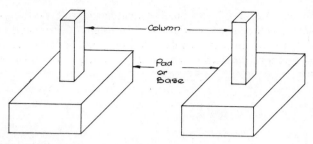

Fig. 6.13 Pad foundation

Where there are a large number of columns in a small area, or where the load carried by a series of columns is such that there is only a small distance between each pad, it may be more economical to construct a slab or raft over the whole area, rather than a series of individual slabs (fig. 6.14).

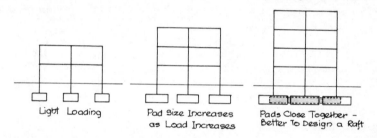

Light Loading

Pad Size Increases as Load Increases

Pads Close Together – Better To Design a Raft

Fig. 6.14 Combined pad foundations

6.6 Foundations to beds and pavements

A bed, in construction terms, is the area upon which a large storage vessel rests, i.e. the base of the vessel. The layer between the natural foundation and the bed must be able to transfer the weight of the vessel and its contents evenly to the natural foundation without undue settlement in any one section. In order to do this, the layer material itself must be free-draining and of uniform consistency, so that moisture or changes in the loading conditions do not cause any change in the volume of material. For this reason, the material selected is generally crushed rock having a uniform particle size of approximately 35 mm, laid and consolidated by rollers to depths varying between 200 and 750 mm (fig. 6.15). The free-draining facility tends to prevent the natural foundation becoming waterlogged and resulting in a rapid loss of bearing strength.

The bed is also that part of the construction which supports pipes, either above or (more usually) below ground (fig. 6.16). In this case it is also essential that no movement takes place, other than that which has been designed for,

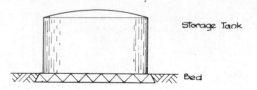

Fig. 6.15 Bed to foundation

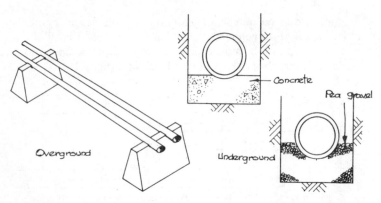

Fig. 6.16 Drain-pipe bed

as this may cause unnecessary distortion or fracture of the pipework. Similar methods of construction to those described above are used, but in the case of underground pipework a smaller size of uniform gravel (20 mm) is placed to a depth of approximately 150 mm under the pipes and acts as a cushion.

Pavements may be considered as any paved area such as roads, footpaths, and drives. They have to be designed to carry loads whose point of application and magnitude vary rapidly. The foundation, or base, must again transfer these loads to the natural foundation or fill, known as the subgrade, without deformation and must also protect the subgrade from excess moisture. The foundation for a road is subjected to higher loadings than that for footpaths and is generally constructed in two layers, the sub-base and the base (fig. 6.17). The base, being closer to the surface, has to carry higher stresses and is, therefore, constructed of higher-grade materials than the sub-base. Both layers should be well compacted during construction, and traffic is frequently allowed to run on the base construction (covered by a temporary surfacing for protection) prior to the final surfacing work being carried out, thus providing additional compaction under working conditions.

The materials used for the sub-base are hardcore or crushed rock; those for the base of the pavements are hardcore, crushed rock, clinker, or pulverised

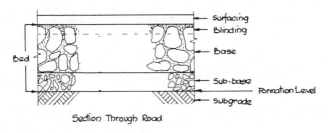

Fig. 6.17 Road bed

fuel ash (p.f.a.). The surface of the base should be relatively smooth in order to receive the pavement surface, and a blinding of sand, quarry waste, or p.f.a. is applied and rolled into the top surface.

7   The basic structure

Understands the primary functions and identifies component parts of basic structure.

7.1   Identifies the superstructure and explains relationship to sub-structure.
7.2   Explains basic concept of a structure as that which transmits forces from points of application to points of support.
7.3   Identifies the basic types of structure.
7.4   Sketches and describes typical structural forms.
7.5   Identifies the component parts of a structure and explains their functions.

Acknowledgement is due to the Technician Education Council for permission to use the content of the TEC units in this chapter. The council reserves the right to amend the content of its units at any time.

7.1 The superstructure

The superstructure, in general, carries the loads which are imposed on a building, in the form of dead and live loading, and transfers them to the substructure.

The dead loading comprises the self weight of primary, secondary, and finished elements, together with the services installations and other fixtures and equipment which are of a permanent nature.

The live loading comprises the weight of the occupants of a building together with their movable fittings and impedimenta; the forces exerted on the structure by wind, rain, snow, and personnel who, together with their equipment, carry out maintenance and repair work; and any other force or load which is applied for only a short period of time or is likely to be moved from one position to another.

The substructure and superstructure are generally designed to operate as one unit, although in construction the completion of the substructure is considered as being the completion of a stage of the work.

7.2 Design of the superstructure

The loadings previously described occur at many points in the superstructure and the design of the superstructure must be such as to safely transfer these loads from their points of application to the points of support, i.e. either the substructure or the foundation.

The simplest example of this load-transference is that of the arch, in which the vertical loads above an opening try to flatten the arch. However, by its form and the shape and strength of the materials used in its construction, the arch transfers those loads to the sides of the opening.

If a number of arches are joined together, a framework of arches and columns will result with each part dependent, to some extent, on its neighbour for support (fig. 7.1). In this way a load-bearing framework may be built up with each component of the structure having a role to play.

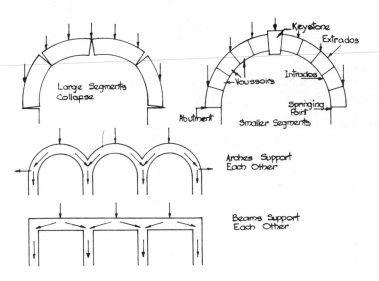

Fig. 7.1 Arches

7.3 Types of structure

There are many types of structure which are used in construction at the present time. Each has been designed or evolved to give a better use of structural materials, namely concrete, steel, timber, or brick; to provide an easier method of construction or erection; to reduce costs; to provide an answer to a particular problem, such as the provision of a clear floor area for warehousing; and, by no means least, to conform with the requirements of an architect in his search for visual appeal.

There are three basic types of structure: (a) solid, (b) skeleton, and (c) surface, each of which uses the strength of a given material and combines it with shape to achieve load-transference. It should be appreciated that one or more of these basic types may be used in the design of a complete building.

a) *The solid structure* combines the load-carrying function with that of space enclosure. The loads are transferred to and spread through the walls to give a distributed load on the substructure; or, in the case of a roof, the load is spread on to the supporting structure underneath.

b) *The skeleton structure* comprises a framework through which the loads are concentrated and transferred, still in concentrated form, to the supporting structure or substructure. In this case, the strength of the members of the framework and their connections play an important part in the transmission of the applied loads.

c) *The surface structure* is one which has a thin skin which is (i) sufficiently rigid to be self-supporting when shaped, the shape providing additional strength, and (ii) sufficiently strong but flexible enough to support load when stretched across a supporting framework or medium.

7.4 Structural forms

There are many structural forms which have been developed from the three basic types.

Solid structures (fig. 7.2)

a) *Cellular*   The loads are dispersed quickly to the 'walls' of the cells, each wall being rigidly jointed to its neighbours. In many cases the thickness of cell walls may be reduced because of this load-sharing ability. The resulting structure is rigid and stable, suited to applications where large areas are not required or alterations in layout are unlikely.

b) *Cross-walls*   A series of independent walls, built at right angles to the front elevation of the building, which carry the floor and roof loads. They are usually built at standardised centres of up to 6 m, thereby allowing other elements also to be standardised. As a free-standing wall is unstable,

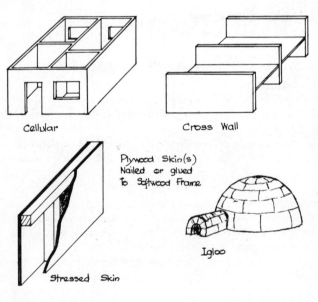

Cellular

Cross Wall

Plywood Skin(s) Nailed or glued to Softwood Frame

Stressed Skin

Igloo

Fig. 7.2 Solid structures

it is essential that lateral ties be created, and this is achieved by the floors being rigidly built into the walls, thus creating a box-type structure. Another means of stabilising the wall is to create a 'T' shape at each end. This form of structure is suitable for buildings of up to five storeys in height.

c) *Stressed skin*   The development of timber housing and the use of timber stressed skins — a softwood framework, covered with plywood which strengthens and is strengthened by the framework — provide another example of a load-bearing 'solid' structural member which is used as a wall or floor component.

d) The *igloo* is a further example of the solid structure, although the circular form is not generally favoured.

Skeleton structures (fig. 7.3)

a) *Apex frame*   The framework of poles is jointed at or near one end to provide the framework of a conical shape, such as in the case of a wigwam.

b) *Rectangular frame*   A series of uprights and cross-members set up in mutually perpendicular planes provides the framework for support to the floors, walls, and roof. This framework is the most common of the skeleton building frames.

c) *Truss*   The truss is based on the triangle, since the triangle is the most rigid of all shapes. The truss is usually found in pitched-roof structures, as its triangular shape is ideally suited to that situation.

Rectangular concrete-frame buildings

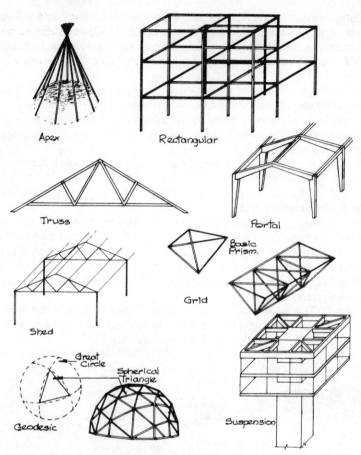

Fig. 7.3 Skeleton structures

d) *Portal frame* Similar to an arch, but consisting of two uprights rigidly jointed by a horizontal, sloping, or curved third member. Each frame requires lateral support, usually in the form of bracing from other similar frames.

e) *Shed frame* Similar to the portal frame but the third member is in the form of a roof truss. As its name implies, it is suitable for long clear-span buildings.

f) *Grid frame* Used for lightweight roof structures covering large open floor areas, the frame comprises a series of triangular frames set out in the form of a grid. The grid may be in the form of one, two, or three layers or of a space grid which comprises a six-member frame joined to other similar frames forming a very strong rigid framework.

g) *Geodesic frame* Formed in the shape of a dome and comprising a network of triangular frames in the form of spherical triangles, i.e. portions of a sphere formed by the intersection of great circles (a great circle has a diameter equal to that of the sphere it is drawn on). The greater the number of triangles used in a structure, the less is the chance of collapse. This frame can be used to enclose large volumes at minimum cost, using standard components.

h) *Suspension frame*   Used in multi-storey construction, it comprises a central solid support structure extending to the full height of the building, at the top of which a rigid horizontal support structure is cantilevered out over the plan area of the building; from this frame the floors of the building are suspended by cables or flat steel bars.

Surface structures (fig. 7.4)

a) *Shell dome* ⎫
b) *Barrel vault* ⎬ Both structures use their curved shape to obtain strength.

c) *Folded or bent plates*   If any thin sheet material is folded or bent it will gain in strength. The steel body of a modern car is a good example of the strength obtained from bending (pressing) a flat steel sheet into some other shape. By folding a piece of paper, its strength and ability to support load is tremendously increased. This fact is used in both wall and roof construction where thin-skin members are used.

d) *Suspension roofs and tents*   In these, the membrane, stretched over some other skeletal frame, itself forms a structural component which will carry wind and other loadings.

e) *Air-supported structures*   In this case the whole skin is supported by the medium of air pressure from the inside — in a similar manner to a balloon.

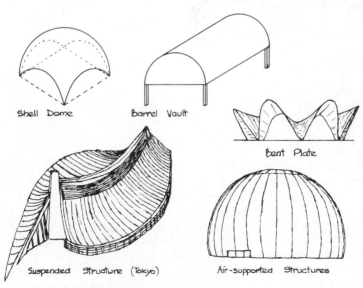

Shell Dome   Barrel Vault   Bent Plate

Suspended Structure (Tokyo)   Air-supported Structures

Fig. 7.4 Surface structures

The edges of the skin must be securely anchored at ground level, and entry is by means of an airlock. This type of structure is suitable for covered sports arenas, temporary warehousing, and even a bad-weather cover for urgent construction operations.

7.5 Structural components

The main components of most frame structures must be able to resist compression, tension, bending, or a combination of all these forces.

a) *The beam* (fig. 7.5)   This is the horizontal member of a framework, and its duty is to transfer the loads imposed on it to the points of support. Most structural members tend to deflect under load (a maximum of $\frac{1}{360}$ of the span is allowable), and this creates tension in the bottom of the beam and compression in the top as a result of the bending taking place. The loading also creates a shearing or cutting action on the beam at or near the supports, or along its length as a result of the bending. Slender beams i.e. those which are deep but not very wide, may buckle in a sideways direction unless that movement is restrained by lateral support.

A beam, in its simplest form, provides support over an opening for the wall above and is known as a lintel. In more complicated rigid frameworks, the beam not only transfers the loads imposed on it to the supports — the columns — but may, as a result of its own bending, also bend the columns.

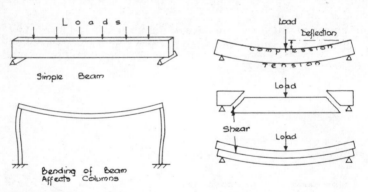

Fig. 7.5 The beam

b) *The column* (fig. 7.6)   A vertical member, generally in compression, which transfers the loading of the beams to the substructure. This member will tend to be either squashed or bent as a result of the loadings applied to it. The squashing or buckling action occurs as a result of the column material being unable to withstand the compressive loading. Any slender member which has loads applied to its ends, such as a column, will tend to bend; this bending tendency will increase with the application of off-centre loads, or because of the bending of beams fixed rigidly to the column. The column must, therefore, be of a suitable shape and made of suitable materials to withstand these effects.

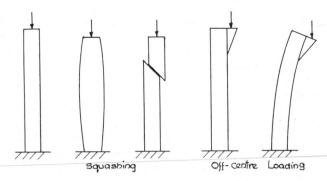

Fig. 7.6 The column

c) *The slab* (fig. 7.7)　Forming a floor or flat roof, the slab is made of concrete and can be considered as a series of beams linked together spanning from one support to another, as in the case of cross-wall construction. The strength of the floor may be increased or its thickness reduced if it is designed to span in two directions, as would be the case in floors formed on the rectangular-frame structures.

　　The slab may be strengthened by beams or it may be used to provide the lateral restraint required where slender beams are used.

d) *The strut* (fig. 7.8)　This is the term used for a member which is only in compression. A strut occurs in a triangulated framework where the load is applied at one end along the axis of the member and is transferred by the member to its other end.

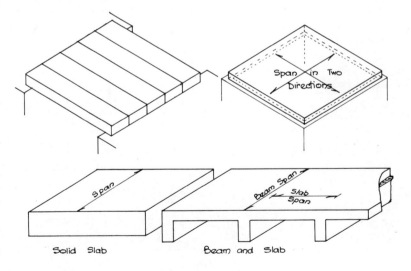

Fig. 7.7 The slab

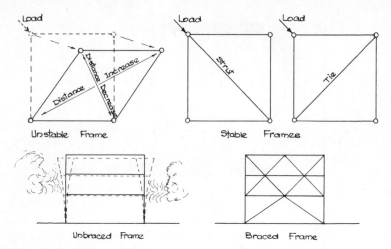

Fig. 7.8 Struts, ties, and braces

e) *The tie* (fig. 7.8)   This is the term used for a member which is in tension along its axis. Ties also occur in triangulated frameworks.

f) *Braces* (fig. 7.8)   These are members which may act as struts or ties, depending on the loading condition. They usually change a rectangular form into one of triangles and are used mainly to resist forces which act sideways, as in the case of wind.

g) *The connections* (fig. 7.9)   The joints which are formed between the various structural members are known as the connections, and they play a very important role since it is through them that the load-transference takes place. There are many methods of forming connections, ranging from the push-fit joint, through nails, screws, and adhesives, to bolts, rivets, and welds.

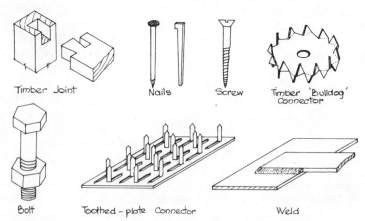

Fig. 7.9 Connections

8   The external envelope

Acknowledgement is due to the Technician Education Council for permission to use the content of the TEC units in this chapter. The council reserves the right to amend the content of its units at any time.

8.1 The elements of the external envelope

Since his earliest days, man has instinctively required a shelter for himself and his possessions. In more recent times he has also provided shelter for many of his work, recreational, and spiritual activities. The function of this shelter is to give protection from the elements, enclose space, and provide a suitable internal environment for the activities taking place within it; at the same time, the shelter should not detract from man's external environment.

The shelter can be described as the 'external envelope' which fulfils the functions already outlined. This envelope is generally comprised of walls and roofs, but, since man requires light and access to his internal environment, windows and doors must also be considered as part of the envelope (fig. 8.1).

8.2 Functions of the external envelope

Having identified the elements which form the external envelope, it is necessary to identify and consider the primary functions of this envelope. The importance of these functions will depend upon the type of structure which is to be erected, the situation of the building, in terms of both macro- and

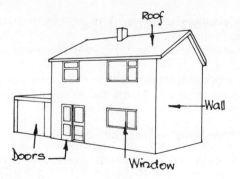

Fig. 8.1 The envelope

micro-environment, and the use for which the building is intended. The functions could be considered under the following headings (see fig. 8.2):

a) strength and stability,
b) weather exclusion,
c) thermal insulation,
d) sound insulation,
e) durability,
f) fire resistance,
g) appearance,
h) feasibility.

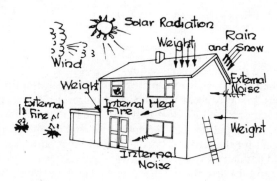

Fig. 8.2 Resistance of envelope

a) *Strength and stability* The envelope has to be strong enough to carry the loads which may be imposed upon it without excessive deformation and to transfer these safely to the structural frame or foundation. The loading may occur in the form of:

   i) the self weight of the envelope,

ii) loads transferred to the envelope from the internal construction, such as floors and cupboards,

iii) containment loads applied internally, as in the case of air-supported structures,

iv) externally applied loads caused either by the environment, in the form of wind, snow, and rain, or by mechanical means in the form of outbuildings, materials stacking, or maintenance work.

b) *Weather exclusion* Keeping the weather out of a building is an extremely important consideration if the internal environment is to remain constant. The size of the problem will depend on the degree to which the building is exposed to the elements, the exposure being assessed in terms of locality, height of building, altitude, and direction and strength of the prevailing winds. Wind pressure can cause draughts through openings and can drive rain and snow through small cracks in the envelope or into the pores of the envelope material.

c) *Thermal insulation* Heat will flow from a high to a lower temperature and, since the temperature outside the envelope in winter is lower than that on the inside, there will be a flow of heat from the building (fig. 8.3). This means that extra heat is required to replace the heat being lost through the envelope in order to maintain a steady internal temperature. If the rate of heat loss from the external envelope can be reduced by means of insulation, savings can be made on both the quantity of heating fuel required and the size of the heating system.

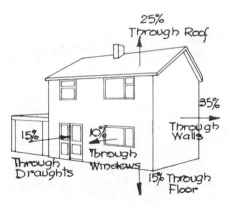

Fig. 8.3 Heat losses

In the summer the envelope should be able to act in reverse, keeping out the heat of the sun and thereby providing a cooler internal environment.

The maintenance of a constant internal environment has been proved to be a prime requirement for optimum performance in both the physical and mental fields of human endeavour.

d) *Sound insulation*   Noise has been defined as unwanted sound. Extremely loud noises can be detrimental to health, while small noises can be irritating and cause loss of concentration. Sound is transferred through both air and material by means of pressure waves and vibrations. When these waves hit the envelope surface, some will be reflected and some absorbed, while others will be apparently transmitted through the envelope.

Since the majority of people do not have control of the sounds outside the building, such as road traffic and aircraft noise, the external envelope must reduce that noise to an acceptable level. It must be remembered that, since sound is airborne, the opening of a window for ventilation or of a door for access or egress will greatly reduce the insulating effect of the overall envelope.

e) *Durability*   The material from which the external envelope is constructed must have sufficient resistance to the damaging effects of the climate – in the form of erosion, atmospheric pollution, frost, rain, and chemical and solar degredation – to provide a building which will be relatively free of maintenance for its anticipated useful life.

f) *Fire resistance*   Many million pounds worth of damage occur to buildings every year through fire. It is, therefore, necessary to keep the effects of fire to a minimum. These effects can be minimised if the building itself can contain the fire until the arrival of fire-fighting appliances.

The prime consideration, in the event of an outbreak of fire, is the safety of human life; hence the external envelope must be strong enough to withstand the ravages of fire for such time as it will take to evacuate the building of all the occupants and allow them to reach places of safety some distance away.

The fire should also be contained within the external envelope for as long as possible to prevent the spread of fire to other buildings in the vicinity.

g) *Appearance*   This is the aesthetic requirement which should be considered at the design stage, so that man may be at one with his artificial environment. The external envelope should, therefore be compatible with others in the immediate vicinity, by either blending with them or providing a contrast which is not too stark. This compatibility can be achieved by considering

i)   the size of the building in both horizontal and vertical planes, e.g. the Empire State building would be just as incongruous in a country village as would be the Tower of London in a new town;

ii) the shape of the building, which should be pleasing both to the client and to the planning authorities;

iii) the colour, surface texture, and type of materials which form the envelope;

iv) the fenestration of the building;

v)   the way in which the building will appear in its future environment and after weather and atmospheric pollution have taken their toll.

In rural areas the envelope may be required to blend with the landscape so that it becomes as inconspicuous as possible.

h) *Feasibility* The foregoing functions of the envelope must not only be considered individually, but also as a whole. Feasibility is the overall consideration of these requirements, having regard to the specific space and enclosure, and the constraints imposed by the limitation of resources, legislation, and current practices.

8.3 Choice of materials for external walls

The materials which are commonly used in external walls are natural and artificial stones, bricks, blocks, concrete, timber, plastics, glass, metal, and asbestos cement. Before the selection of a material for a wall or any other component can be made, an understanding of the composition, advantages, and disadvantages of these materials is required.

The factors which affect the choice of materials for an external wall are:
a) the structural form of the building,
b) the type of wall required,
c) availability,
d) speed of erection,
e) cost.

a) The structural form of the building determines whether a wall is to be load-bearing or non-load-bearing; hence the nature of the loading and the load-carrying capacity of the material will restrict the choice to certain materials.
b) Emphasis on particular functions of the external envelope will provide a constraint on the designer, limiting his choice of materials as well as of wall construction, e.g. an isolated single-storey building in an exposed situation will require emphasis on weather exclusion, thermal insulation, and durability, while low on the order of priorities would be sound insulation and fire resistance.
c) The selection of some special material may mean that it is available from only one supplier and there may be a long wait before an order can be fulfilled. On the other hand, the national economic climate may mean that certain of the more common materials, which are under normal circumstances readily available, may be in short supply.
d) The speed of erection of the wall may be of prime importance if the building is urgently required.

Materials which are associated with wet trades, i.e. brickwork and in-situ concrete, will take longer to erect than 'dry' constructions, i.e. timber and large precast concrete units.
e) The cost of a wall is not only the cost of the materials which make up the wall, but also the cost of the labour to erect the wall and any subsequent finishing costs. Hence a wall material which initially appears to be inexpensive may be costly because it is a labour-intensive operation to erect and finish the wall using that material (fig. 8.4).

The suitability of materials must take into account not only the foregoing factors, but also the functions which the external envelope is required to perform; hence every case must be treated on its merits.

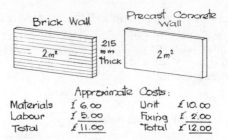

Brick Wall		Precast Concrete Wall
	215 mm thick	
2 m²		2 m²

Approximate Costs:

Materials	£ 6.00	Unit	£ 10.00
Labour	£ 5.00	Fixing	£ 2.00
Total	£ 11.00	Total	£ 12.00

Fig. 8.4 Cost comparison

8.4 Construction of external walls

Solid walls

Solid walls, as the name implies, are walls which other than for openings, have no void deliberately left in them, e.g. for insulation purposes. These walls prevent moisture penetrating to their inside face by either the sponge principle or the impervious-skin principle (fig. 8.5).

a) *The sponge principle*   The wall is of sufficient thickness to prevent moisture which has been absorbed into the pores of the material reaching the inner surface of the wall before evaporation from the external surface draws out that moisture from the pores.

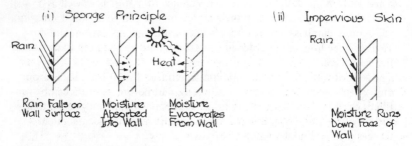

(i) Sponge Principle   (ii) Impervious Skin

Rain

Heat

Rain

Rain Falls on Wall Surface   Moisture Absorbed Into Wall   Moisture Evaporates From Wall   Moisture Runs Down Face of Wall

Fig. 8.5 Moisture resistance

b) *The impervious-skin principle*   The wall has an external surface with either a self or an applied finish which does not readily allow moisture to pass through it into the remainder of the wall.

Being constructed in a dense form, these walls provide good sound insulation, but allow heat to pass through.

The solid wall is generally constructed of regular-sized rectangular blocks so that, when these are laid in layers (known as courses), there are continuous horizontal joints. However, in order to ensure that any forces or loads which are applied to the wall are quickly spread throughout the wall, and to increase the wall's stability, continuous vertical joints should be avoided.

The method of avoiding the continuous vertical joint is known as bonding (fig. 8.6).

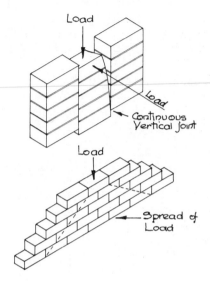

Fig. 8.6 Reason for bonding

The standard brick size is 215 mm x 102.5 mm x 65 mm, and to avoid confusion when referring to brick arrangements the parts of the brick are given different names (fig. 8.7). Some bricks have no frogs or indents, some have two, and others may have holes of various shapes passing through them from top to underside. The size of the brick is such that, with allowance for jointing, two header faces equal one stretcher face and three course heights also equal one stretcher face (fig. 8.8).

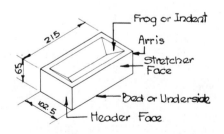

Fig. 8.7 Brick terminology

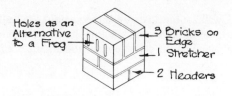

Fig. 8.8 Relationship of brick sizes

Stretcher bond   A wall which is to be 102.5 mm thick is built with the stretcher faces of one course overlapping the stretcher faces of the course below by half the face length, so that a regular pattern is built up (fig. 8.9).

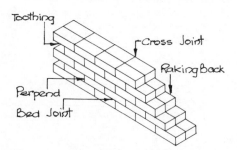

Fig. 8.9 Stretcher bond, including brickwork terms

Header bond   Where a wall is required to carry heavier loads or to keep out the rain, it will have to be thicker than 102.5 mm. This can be achieved by laying the bricks with the header faces showing, each layer overlapping the previous layer by half the horizontal face dimension, thus giving a wall of 215 mm thickness (fig. 8.10).

Fig. 8.10 Header bond

Wall thickness is usually denoted in terms of the stretcher face length (one brick length), which is abbreviated to 'B'. Hence the 102.5 mm thick wall is $\frac{1}{2}$B thick, the 215 mm thick wall is 1B thick, and so on in $\frac{1}{2}$B increments.

The visual appearance of a wall in header bond is unattractive, and a 1B wall in stretcher bond would give a continuous vertical joint in the middle, which is a weakness in the bond. In order to overcome these problems, other bonds using combinations of headers and stretchers were devised.

English bond   The wall is constructed using alternating header courses and stretcher courses, giving a strong bond (fig. 8.11).

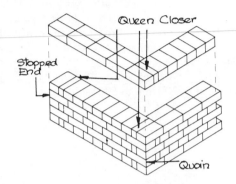

Fig. 8.11 English bond with corner and stopped end

Flemish bond   The wall is constructed using alternate headers and stretchers in each course (fig. 8.12), but it is more economical than English bond since there are less facing bricks required.

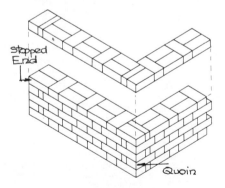

Fig. 8.12 Flemish bond with corner and stopped end

Where a wall has to be stopped, i.e. at an opening or a corner, the bond must be maintained, for both structural and aesthetic reasons. This may require the cutting of bricks, and these cut bricks, inserted in a wall half a brick length from the corner (quoin) or end, are called closers.

Stone walls Natural stone, although durable, is expensive, especially when cut to regular shapes, and is nowadays used mainly as a facing to provide a feature or to blend with the surroundings.

However, there are several types of solid stone walling which are still seen in old buildings or in country districts where the stone is readily available (fig. 8.13).

a) *Uncoursed random rubble* The stones are of random size and shape and laid in such a way that there are no continuous horizontal or vertical joints.

b) *Coursed random rubble* Similar to (a) but horizontal joints are present at intervals of 200 mm to 500 mm throughout the height of the wall.

c) *Uncoursed square rubble* Similar to (a) but the stones, still in random sizes, have been squared off.

d) *Coursed square rubble* A combination of (b) and (c) where squared stones are laid in courses 300 mm to 500 mm thick.

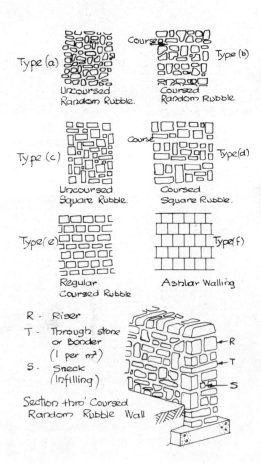

Fig. 8.13 Stone-wall details

e) *Regular coursed rubble*   Squared stones of the same thickness are layed in courses.

f) *Ashlar walling*   Square stones having a worked face are used in random or coursed construction, usually with a brickwork backing.

Uncoursed square rubble wall with dry joints

Coursed square rubble wall with mortar joints

Jointing   The jointing of all brick and stone walls is by means of mortar. Mortar is a mixture of sand, lime, cement, and water to provide the chemical reaction for setting to take place. Ideally the mortar should (i) be workable enough to give ease of laying, (ii) bond the bricks or stones together, (iii) fulfil the functions required of the wall.

The mortar should be slightly weaker than the bricks so that, if any movement of the wall takes place, cracking will occur in the mortar rather than in the bricks. This cracking is easily made good, whereas if the bricks cracked there would be a weakening of the structure.

Typical mortar mixes (by volume) are

Cement mortar	1:3	(1 part cement to 3 parts sand) Suitable for exposed brickwork.
Gauged mortar	1:1:6	(cement: lime: sand) Suitable for most conditions.
	1:2:9	Suitable for most conditions with the exception of severe exposure.
	1:3:12	Suitable for internal use only.
Lime mortar	1:3	(lime: sand) Suitable for internal use only.

Lime is included in the mix to improve the workability of the mortar, but more often in modern construction an additive known as a plasticiser is used.

The face of the joint may be finished in a number of ways, this finishing work being carried out before the mortar has finally set. It has the effect of creating a denser surface to the mortar, thus improving the weather resistance of the joint. The joint finish also adds to the overall aesthetic appearance of the work (fig. 8.14).

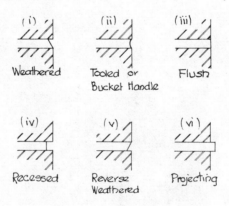

Fig. 8.14 Methods of jointing (N.B. Types (iv), (v), and (vi) are not recommended for exposed situations since water will lodge on the horizontal face.)

Cavity walls

In order to provide adequate resistance to the penetration of moisture in most parts of Britain, a solid external wall would need to have a thickness of at least 350 mm (over 1½B thick), and to provide adequate thermal insulation it would need to be at least 450 mm (over 2B) thick. The cost of constructing the external walls can be reduced by introducing an air space or cavity between the outside and inside faces of the wall. This cavity prevents the passage of moisture through the wall and also improves its thermal insulation.

The cavity wall is constructed by building two ½B thick walls, or leaves, in stretcher bond, with a 50 mm cavity between them (fig. 8.15). This form of construction should not exceed 7.6 m in height because the wall will become

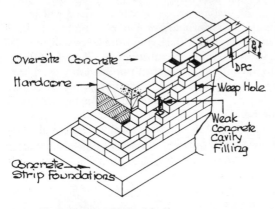

Fig. 8.15 Cavity wall with corner racked back

unstable. The stability, however, is improved by 'tying' the two leaves together, and this is achieved by placing wall ties at intervals across the cavity and bedding them in the horizontal joints of each skin (fig. 8.16). The ties should be designed to prevent the passage of moisture across them (fig. 8.17).

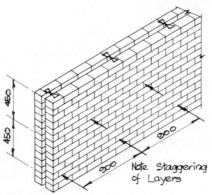

Fig. 8.16 Positioning of wall ties

93

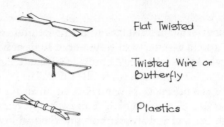

Fig. 8.17 Types of wall tie

In this form of construction it is essential that the cavity should be kept absolutely clear, otherwise moisture may bridge the gap. The gap is generally bridged only in the case of window or door construction or below the level of the damp-proof course, where the cavity is frequently filled with concrete to provide additional strength to the substructure. The top of the concrete filling should slope to the outer skin, which in turn has every third vertical joint of the course immediately above the filling left open to provide weep holes. The filling and the holes thus direct any moisture dropping from the cavity ties to the outside ground.

In order to reduce the loss of heat through external walls to amounts prescribed in the Building Regulations, the inner leaf of the cavity wall may be constructed with insulation blocks (fig. 8.18).

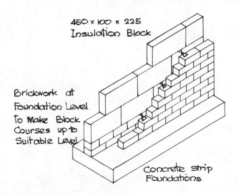

Fig. 8.18 Cavity wall with block inner leaf

Damp-proof course

The upward movement of moisture by capillary action in the material of the walls, in contact with the subsoil, must be prevented; otherwise the internal wall finishes and other items in direct contact with the wall will be detrimentally affected. An impermeable barrier is, therefore, inserted in the wall to stop rising moisture at a height not less than 150 mm above the finished surface of the ground in an external wall. The more commonly used materials for a d.p.c. include bitumen with a hessian, fibre, or asbestos base; polythene; mastic asphalt; two courses of engineering bricks in 1:3 cement mortar; or slate.

8.5 Requirements of windows

The primary functions of a window are to provide (i) natural lighting inside the envelope and (ii) a means of ventilation. In terms of performance requirements, the window, forming part of the external envelope, must provide resistance to (i) wind pressure, (ii) water penetration, and (iii) air penetration in the form of draughts. It must also fit into the space allocated in the external envelope.

Wind pressure on the outside face of a window will cause both glass and frame to deflect, and the amount of this deflection must be limited if breakage is to be avoided. The limitation can be by size of pane, thickness of frame, or thickness of glass. Wind pressures are graded from very well protected to severe (see fig. 8.19) and graphs can be prepared which limit the size of glass in respect of these pressures (see fig. 8.20).

Severe	Up to 2300 N/m²
Moderate	Up to 1900 N/m²
Sheltered	Up to 1500 N/m²
Well protected	Up to 1200 N/m²
Very well protected	Up to 900 N/m²

Fig. 8.19 Design wind loadings

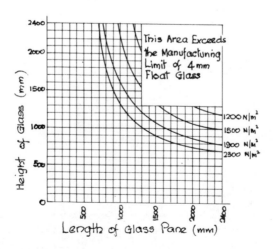

Fig. 8.20 4 mm float-glass exposure-limit graph

If a window has a section which will open to provide a means of ventilation, the joint between the opening and fixed sections will be a weak point for the entry of rain and air. BS 4315:Part 1:1968 establishes methods of testing the resistance of windows to air and water penetration, and BS.DD4:1971 gives minimum performance standards for water penetration and air infiltration.

Windows are now manufactured in standardised sizes which are designed to fit into modular openings, with the minimum of gap between window and wall.

8.6 Types of window

Windows may be classified either by the material from which the framework is constructed — i.e. timber, steel, aluminium, plastics — or by the method of opening — i.e. hinged, pivoted, or sliding (fig. 8.21). Those windows which

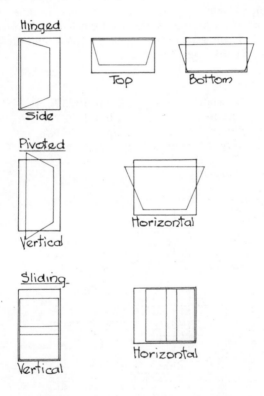

Fig. 8.21 Window classification

have hinged opening lights are known as *casement* windows, and those having pivoted or sliding lights are known as *sash* windows (a light in this context being a single glazed unit of a window).

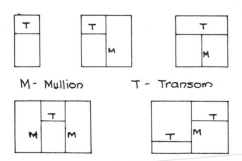

Fig. 8.22 Mullions and transoms

8.7 Window parts and fittings

A casement window has two main parts: the frame and the casement. The frame comprises a *head* and *sill* which span the full width of the frame; *jambs*, which are the vertical side members; *mullions*, which are vertical members of the frame other than the jambs; and *transoms*, which are horizontal members of the frame other than the head or sill (fig. 8.22).

The casement (figs 8.23 to 8.25) comprises a *top rail*, a *bottom rail*, and *stiles*. Note that the side of the casement which is hinged is indicated by the apex of the arrow drawn on the window.

The terminology for metal windows remains the same as for timber windows, although the shape and profile of the framing members is somewhat different (fig. 8.26).

There are three basic pieces of ironmongery required for opening sashes (fig. 8.27): the hinges, made out of steel or brass; the fastener, for security; and the stay, which keeps the sash in position once open. The fastener and stay, usually a matched pair, are made out of steel, aluminium, plastics, or wrought iron.

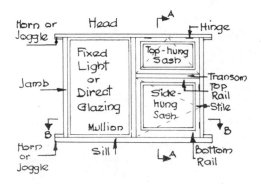

Fig. 8.23 Casement-window terminology

97

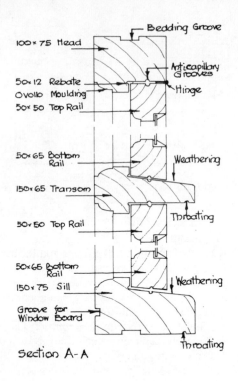

100×75 Head

Bedding Groove

50×12 Rebate

Anticapillary Grooves

Ovollo Moulding

Hinge

50×50 Top Rail

50×65 Bottom Rail

Weathering

150×65 Transom

50×50 Top Rail

Throating

50×65 Bottom Rail

Weathering

150×75 Sill

Groove for Window Board

Throating

Section A-A

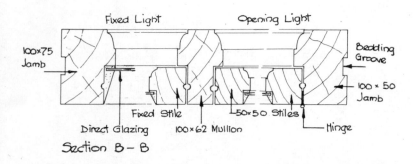

Fixed Light

Opening Light

100×75 Jamb

Bedding Groove

100×50 Jamb

Direct Glazing

Fixed Stile

Fixed Stile

100×62 Mullion

50×50 Stiles

Hinge

Section B-B

Fig. 8.24 Traditional timber casement (note alternative fixed-light construction)

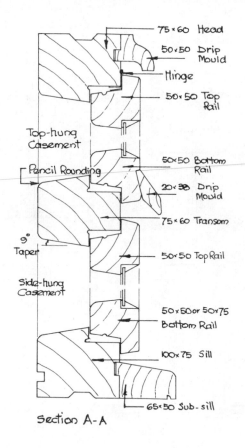

75×60 Head

50×50 Drip Mould

Hinge

50×50 Top Rail

Top-hung Casement

Pencil Rounding

50×50 Bottom Rail

20×38 Drip Mould

75×60 Transom

9° Taper

50×50 Top Rail

Side-hung Casement

50×50 or 50×75 Bottom Rail

100×75 Sill

65×50 Sub-sill

Section A-A

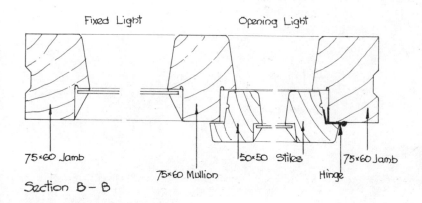

Fixed Light   Opening Light

75×60 Jamb

75×60 Mullion

50×50 Stiles

Hinge

75×60 Jamb

Section B - B

Fig. 8.25 Modified BS timber casement

99

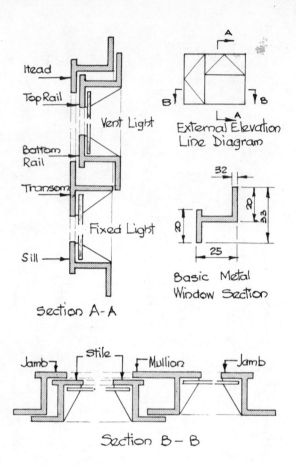

Fig. 8.26 Details of metal windows

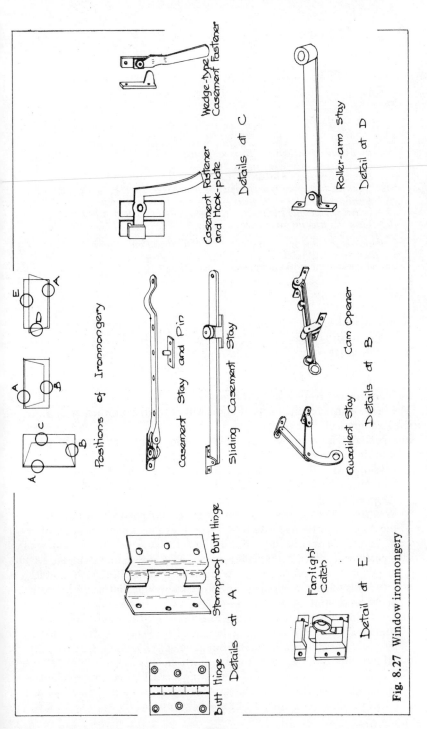

Fig. 8.27 Window ironmongery

101

9   Doors

Understands the functions of doors and fundamentals of their construction.

9.1 Identifies typical performance requirements of doors.
9.2 Identifies basic door types and names the members in their fabrication.
9.3 Identifies typical sizes of doors, door sets and door openings.
9.4 Draws typical door frame assemblies and names the component parts.
9.5 Identifies relevant ironmongery.

Acknowledgement is due to the Technician Education Council for permission to use the content of the TEC units in this chapter. The council reserves the right to amend the content of its units at any time.

A door is a movable barrier element used to seal an opening in the external envelope or between sections within the envelope. The primary function is to allow the passage of people, vehicles, and goods from one area to another, while maintaining the various performance requirements.

9.1 Performance requirements of doors
The performance requirements of doors will depend upon whether they are situated internally or externally, as follows.
a) *Weather exclusion*   An external door, being part of the envelope, must provide the same degree of weather exclusion as the remainder of the envelope. As with the windows, the joint between the fixed and opening sections of the door is the weak point, therefore in areas of severe exposure the positioning of the door may be critical.
b) *Security*   The security of a door depends on the materials from which the door is constructed, and BS 459 specifies a minimum thickness of 44 mm for an external door in order to deter burglars. The positioning of hinges, locks, bolts, letter-plates, and windows is also important and recommendations covering these aspects are made in BRS Digest 122. Internal security is seldom required in the domestic dwelling, but may be necessary in offices or storerooms.
c) *Fire resistance*   The door, acting as a barrier element, must be able to contain fire, thereby minimising the spread and hence the damage caused by a fire. The Building Regulations require large buildings to be sectioned off into fire-resisting compartments which should be able to contain a fire for a given period of time. Doors allowing access between these compartments must, therefore, have the same resistance to a fire as the remainder of the compartment and must be specially designed.

d) *Thermal and sound insulation*   The size of a door opening in relation to the size of the surrounding wall area is small and, unless there is a special need for insulation, e.g. in a cold store, the door is not specifically designed to fulfil these requirements.

e) *Privacy*   This is achieved immediately an unglazed door is placed in an opening, but, by careful planning of a layout, privacy may also be achieved when a door is open, e.g. at entrances to toilets. Partial privacy may be obtained by the use of obscured rather than clear glass as a means of giving additional light to a doorway area.

f) *Operation*   The operating conditions of a door, its size, and its position in a building will determine the method of opening that door. A door for pedestrians may be simply hinged, but this method would be unsuitable for a large industrial door. The size and weight of a door may determine whether hand or electrically operated gearing is required to open it.

g) *Durability*   A door must be able to withstand the use for which it is intended, and this means that the materials from which the door is made must be of adequate strength, and be properly joined together to form a rigid article. The materials should be dimensionally stable under the full range of operating conditions — flush timber doors will twist when subjected to differences in humidity and temperature between the two faces. The durability of a door may be increased by the selection and use of the correct ironmongery, e.g. kicking- and finger-plates.

9.2 Door types and members

Definitions (fig. 9.1)

Door	A movable screen across an opening, providing access to a building or between rooms within a building.
Fanlight	A fixed or opening glazed unit in the framework above a door.
Frame	The surround to a door opening which also supports the door, or the surround to the door itself.
Head	The horizontal member of the frame above the door.
Jamb (post)	The vertical member of the door frame.
Lining	The trim around a door opening.
Mortice	A slot or hole to receive a lock, or forming part of a timber joint.
Muntin	An intermediate vertical member of the door or frame.
Rail	A horizontal member of the door framing. Locking rail — provided to accommodate the lock.
Sill (cill)	The horizontal member of the frame below the door.
Stile	An outer vertical member of the door framing. Hanging stile — the stile by which the door is hung. Closing (locking/striking) stile — the stile which closes against the jamb of the surrounding frame. Meeting stile — the stile which closes against another stile in a pair of doors.

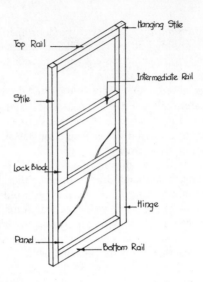

Fig. 9.1 Door terminology

Threshold	See 'Sill'.
Water(weather)-bar	A metallic strip under the sill, preventing moisture penetration.
Weather-board	A horizontal member fixed to the outside face of an external door to prevent weather falling directly on to the sill.

The basic door types are classified by their method of construction, being either unframed or framed, and by their type of opening. Details of these are set out in the various parts of BS 459.

a) Unframed door types

i) *Ledged and battened*   An inexpensive door used in situations where strength is not required and there will be low usage; prolonged use will cause the door to distort or drop. The battens, which are tongued-and-grooved boards, are fixed vertically on to the rigid backing by means of nails, and the battened facing, known as the match boarding, usually has the long edge chamfered so that the joint between battens forms a feature.

ii) *Ledged, braced, and battened* (fig. 9.2)   Similar to the ledged-and-battened type, but a stronger door suitable for out-buildings. The brace slopes upwards between the rails from the hanging edge, to prevent the door drooping.

104

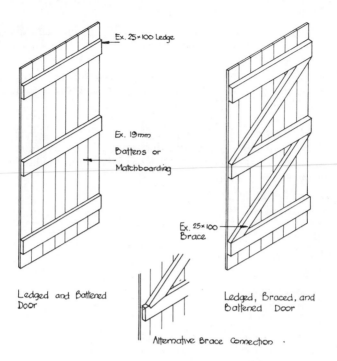

Fig. 9.2 Yard doors

iii) *Glass* Toughened plate-glass doors for shops and offices are supported and hung from metal top and bottom rails. To avoid accidents caused by people walking into the doors, it is common practice to incorporate some form of feature handle with the door.

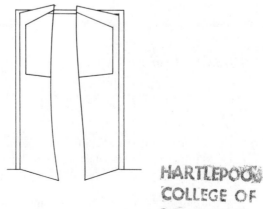

Fig. 9.3 Flexible doors

HARTLEPOOL
COLLEGE OF
29 JUN 1978
FURTHER EDUCATION
LIBRARY

iv) *Flexible* (fig. 9.3)   Self-closing doors used in industrial and hospital situations, where it is not convenient for the user to open and close the door in the normal way. The door comprises a top rail and stile supporting a flexible rubber or plastics sheet, which should be transparent or incorporate a flexible vision panel, in order to avoid accidents or collisions.

b) Framed door types
  i) *Framed and ledged* (fig. 9.4)   Comprises a framework of rails and stiles which gives strength to the door edges, but in conditions of heavy use may tend to drop.

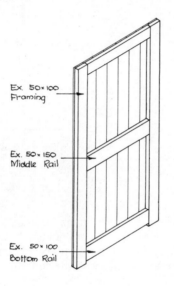

Ex. 50×100 Framing

Ex. 50×150 Middle Rail

Ex. 50×100 Bottom Rail

Fig. 9.4 Framed, ledged, battened door

  ii) *Framed, ledged, and braced* (fig. 9.5)   A more robust door having good weather resistance.
 iii) *Panelled* (fig. 9.6)   There are various arrangements of the framing members, which all have a similar cross-section. The areas between the members are filled in with panels which are fitted into grooves in the edges of the framing, and the panelling may be of glass, timber, or plastics.
 iv) *Flush*   This is the most common type of door used in modern construction and, as the name implies, has two plane surfaces which conceal its internal structure or core. The materials used and the method of construction vary widely, depending upon the situation and function of the door.

   The core of flush doors may be solid, semi-solid, skeleton-framed, or cellular.

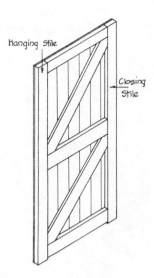

Fig. 9.5 Framed, ledged, braced, and battened door

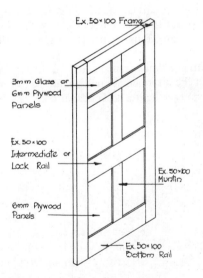

Fig. 9.6 Six-panel door

The solid core (fig. 9.7) is formed either with sheets of chipboard or cork, or from laminated timber which has been planed to size, then butt-jointed and glued. This core provides strength together with good sound insulation.

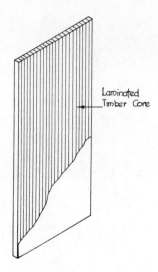

Fig. 9.7 Solid-core door

The semi-solid core is cheaper and lighter in weight than the solid core and comprises an outer timber frame with intermediate horizontal rails, not more than 63 mm apart. The whole core should contain at least 50% timber.

The skeleton or timber-railed core (fig. 9.8) is similar to the semi-solid core but the intermediate rails should be not more than 125 mm apart, giving an even lighter and cheaper door.

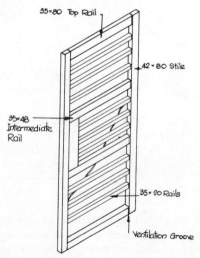

Fig. 9.8 Skeleton-core flush door

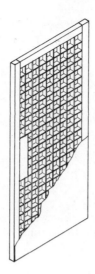

Fig. 9.9 Cellular-core door

The cellular core (fig. 9.9) is formed by the creation of a lattice-work of hardboard, paper board, or cardboard strips, or by the use of closely packed spiral timber shavings. The aim of the core is to provide a low-cost low-weight infill which supports the surface finish and does not allow undulation or rippling.

In the case of semi-solid, skeleton, and cellular-core doors, it is essential that a ventilation channel is formed in the framework to allow free circulation of air between the enclosed spaces in the door and the surrounding atmosphere. At the same time, provision must be made for door locks, the fixing of kicking-plates, and, where required, letter-plates.

The surface finish to a flush door may be of veneer, plywood, hardboard, or plastics.

The veneers, usually of hardwood and having a well-defined grain, are a suitable facing for the solid-core door and provide an excellent finished appearance.

The plywood and hardboard finishes are generally bonded to cores other than the solid by an adhesive such as synthetic resin to BS 1204 or cold-setting casein glue to BS 1444. Plywood is suitable for use on external doors provided that the resin which bonds the plys is weather- and boil-proof (WB P) to BS 1203. A cheaper hardboard facing should conform to BS 1142 and is suitable only for internal doors.

A hardwood edging or lipping strip is provided on the vertical edges of doors in order to protect the edges of the facing material (fig. 9.10).

v) *Glazed* A glass panel or a number of panels can be inserted in both panelled and flush doors. In the case of a panelled door, the construction of the door framework hardly changes, but in the case of flush

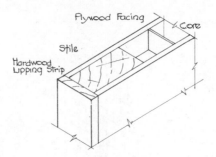

Fig. 9.10 Door edging

doors provision must be made within the core for the framework to receive the glass panel. The inclusion of glazing in a door enables approaching people to see people coming from the opposite direction.

c) Method of opening (fig. 9.11)
  i) *Hinged*   The door is supported on hinges which are fixed to the hanging stile and the door frame; this is known as a side-hung door. The door may open or swing in one or two directions, known as single or double swing, and there may be more than one opening door, or leaf, to an opening.
  ii) *Folding*   A folding door has a number of leaves joined by hinges and may be self-supporting from the frame, but it is more common for the door to be supported by tracks positioned either below or above the door, or by a combination of both. These doors are best suited to large openings where the open door requires less storage space than the sliding door.
  iii) *Sliding*   The doors are individually supported on track which may carry one or more leaves. To facilitate storage, a curved track may be used, in which case the leaves are hinged together. Another form of sliding door is the 'up-and-over' door which is used on domestic garages.
  iv) *Revolving*   The object of this type is to provide a draught-proof opening in the external envelope.

9.3 Sizes of doors, openings, and door sets
Standard sizes for particular doors are given in BS 4787, and the volume and size of traffic passing through an opening will govern the size of the door. Typical domestic door sizes are

Width (mm)	*Thickness* (mm)	*Suitable for*
826	40 or 50	External doors
726	40 or 50	Internal doors
626	40	Cloakrooms and large cupboards
526	40	Broom cupboards

110

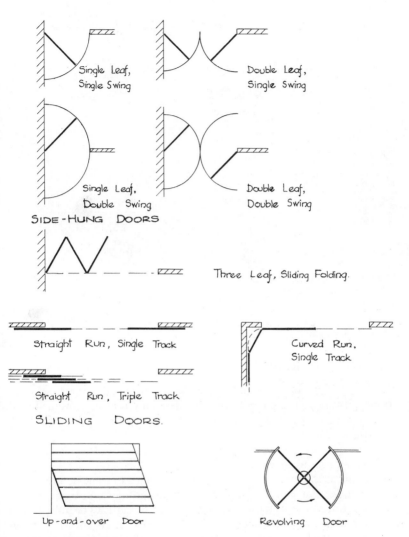

Fig. 9.11 Methods of opening doors

Doors for industrial purposes are usually made to measure or are offered in standardised sizes by specialist manufacturers.

The size of opening provided in a wall to accommodate a door must take into account the thickness of the door frame. In order to accommodate an opening in the overall layout of a building, the dimensions of that opening should be of modular size, i.e. multiples of 300 mm or 100 mm. The standard sizes of opening provided in a wall for doors are therefore 2100 mm × 900 mm,

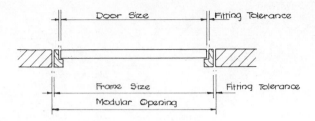

Fig. 9.12 Door sizing

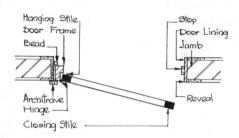

Fig. 9.13 Door-opening terminology

800 mm, 700 mm, or 600 mm. It should be appreciated that the door opening, having been sized by traffic considerations, is finally governed by the size of the modular opening, frame members, and fitting tolerances for the door (fig. 9.12).

The operation of inserting a door in an opening can be time-consuming when a joiner has to do all the work himself. Modern mass-production methods, in both timber and metal door construction, have led to manufacturers being able to provide standard door sets, these sets comprising door, frame, architrave, and ironmongery (fig. 9.14). The door can be partially or fully finished, thereby reducing decorating costs, and will have standard or special ironmongery already fixed before delivery. The door can also be supplied hung in its frame, the hinges being the lift-off type, thereby allowing the door to be removed and stored for protection during the remainder of the construction process. The frame, already made up with or without a sill, can be easily and quickly fixed in the opening, provided the opening is the correct size. The use of door sets can, therefore, save the builder a considerable amount of time compared with the traditional method of construction.

Door sets are manufactured not only with the frame around the door, but also with a framework which extends to the ceiling above the door and which may incorporate a fanlight. This is known as a storey-height frame. The dimensions of all door sets are covered in BS 4787, and typical dimensions of openings are

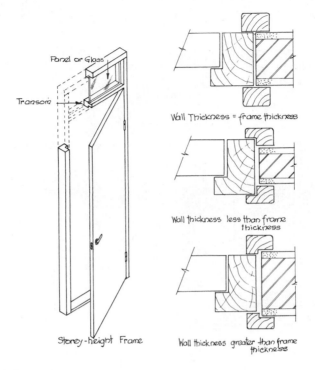

Labels on figure:
- Panel or Glass
- Transom
- Storey-height Frame
- Wall Thickness = frame thickness
- Wall thickness less than frame thickness
- Wall thickness greater than frame thickness

Fig. 9.14 Door sets

	Height (mm)		*Width* (mm)			
	Door	*Storey*	*Single-leaf*		*Double-leaf*	
External	2100	2300 2400	900 1000		1200 1500	
		2700 3000			1800 2100	
Internal	2100	2300 2400	600 700		800 900	
		2700 3000	800 900		1000 1200	
					1500 1800	
			1000		2100	

9.4 Door frames

Door frames serve three purposes: to give a suitable finish to an opening in a wall, to provide a fixing from which the door can be hung, and to provide a suitable joint between the door and the surrounds.

There are two methods of providing a suitably finished door opening: the first method is that of a door lining which covers the whole of the inside of the opening, i.e. soffit and reveals (fig. 9.15), and the second method is that of a door frame which fits within the opening purely as a support for the door (fig. 9.16).

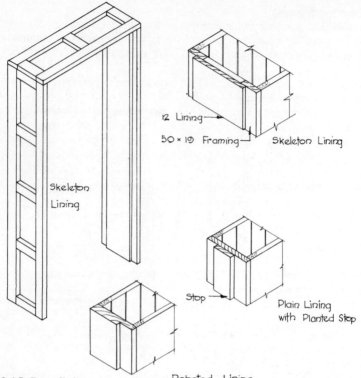

12 Lining

50 × 19 Framing

Skeleton Lining

Skeleton Lining

Stop

Plain Lining with Planted Stop

Fig. 9.15 Door linings

Rebated Lining

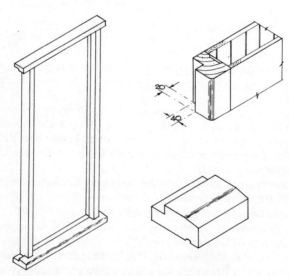

20

40

Fig. 9.16 Door frames

9.5 Door ironmongery

There are two basic pieces of ironmongery required for doors: the hinges and the locking mechanism. Other items which are used in particular circumstances include letter-plates, self-closing mechanisms, and draught-excluders (figs 9.17 and 9.18).

a) *Strap or hook-and-eye hinge*   Made from wrought iron about 50 mm x 6 mm in cross-section and up to 900 mm long, it is screwed and bolted to the door and is a strong hinge used for external doors to garages etc.

   The hook or pintle for gate hinges is of three types: the first is for screwing to a wood frame, the second has a threaded bolt passing right through the post, and the third has a flattened and split bolt for building into brickwork.

b) *Cross garnet or Scotch tee hinge*   Made from wrought iron or japanned pressed steel, it varies in length of blade from 150–500 mm and is screwed to the face of door and frame. It is used for external WC and fuel-store doors and side gates.

c) *Butt hinge*   May be of pressed steel, wrought iron, cast iron, brass, or bronze. Brass butts vary in size from about 25 mm high x 13 mm wide, and they usually have one or two steel washers between each section to take the wear. Iron and steel butts are obtainable from 65–150 mm long (75 mm for windows and 100 mm for internal and external doors in domestic buildings). For heavy doors, three hinges ($1\frac{1}{2}$ pairs) may be used, or one pair of 130 mm butts. A ball-bearing butt hinge is also made for use with heavy doors, and the leaves of the hinge are let into the frame and edge of the door stile, secured by screws.

d) *Rising-butt hinge*   This type of hinge enables a door to be lifted off the frame at will, renders it self-closing, and enables it to clear a carpet when open, yet giving a close joint to the floor when shut. They are made of cast iron, brass, and bronze.

e) *Barrel bolt and tower bolt*   Both these are made in various sizes and metals, and infinite variations of details. The essential difference is that the barrel bolt is wholly encased, whereas the tower bolt runs through rings. The tower bolt is also better for heavy-duty work; the barrel bolt for high-class work.

f) *Mortice lock*   In its usual form it consists of a two- or three-piece wrought-iron or pressed-steel case some 150 mm long, 75 mm high, and about 15 mm thick, with two bronze or brass bolts – one a handle-operated spring bolt and the other a key-operated dead bolt. Lever handles or knob furniture complete the assembly.

g) *Upright mortice or sash lock*   Similar to the mortice lock, this is used for single-panel and glazed doors.

h) *Mortice night latch*   Has one spring bolt, operated by a handle on one side of the door and by a key on the other, and is usually employed for entrance doors.

j) *Rim lock*   This is a lock screwed to the face of the door, as distinct from the inset mortice lock, and is usually used where a door is too thin to accommodate a mortice. The case is usually of japanned steel, but the

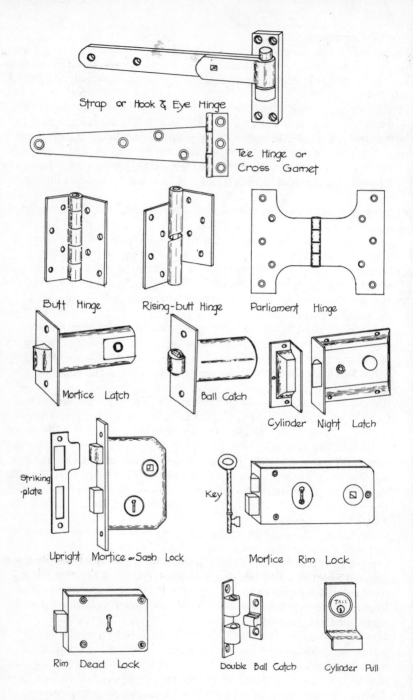

Fig. 9.17 Door ironmongery

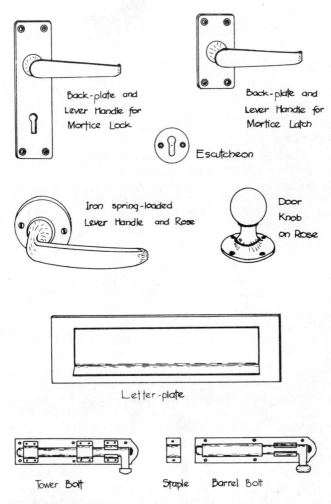

Fig. 9.18 Door ironmongery

better quality may be in wrought iron or brass, and the size is usually about 150 mm x 100 mm. Rim latches and mortice latches are also available – these are similar to the locks, but the term 'latch' denotes that they have only handle-operated spring bolts and not a key-operated bolt.

k) *Rim dead lock*　A dead lock has only one bolt, and this is a 'dead' bolt, i.e. not a spring bolt. It is operated only by a key and the end of the bolt is cut square and not splayed off.

l) *Cylinder night latch*　A cylinder lock is one of the 'Yale' type, the essential parts being a fixed spindle and a rotating cylinder which can only be moved from the outside of the door by a key having the appropriate serrations on its edge, or by a handle on the inside of the door.

10   Roofs

Understand basic roof forms.

10.1   Identifies typical performance requirements for roofs.
10.2   Sketches and describes basic roof forms.

Acknowledgement is due to the Technician Education Council for permission to use the content of the TEC units in this chapter. The council reserves the right to amend the content of its units at any time.

The roof is that part of the external envelope which spans the external walls at their highest level and, being part of the envelope, it must fulfil the functions required of the same (see Chapter 8).

10.1 Performance requirements of roofs
The performance requirements for roofs will vary depending upon the type of building, roof form, and level. These requirements are as follows.
a) *Weather exclusion*   It is essential that any building should have a roof which is weatherproof, as without that protection overhead the occupants of the building might just as well live in the open air. The roof is the first part of the superstructure to be affected by the natural elements — i.e. rain, snow, sun, wind, and dust — and the roof covering should be adequate to resist these effects of nature.
b) *Structural strength and stability*   Supporting the roof coverings there must be some form of framework. This framework must be strong enough to support not only the weight of the coverings and its own weight, but also the superimposed loadings caused by the weight of rain or snow resting on a roof, the weight of workmen carrying out maintenance operations, water-storage tanks situated on or within the roof, or the effects of wind. Wind loading may create a pressure or a suction on a roof, depending on the roof shape and the direction of the wind; the roof must, therefore, be sufficiently stable to resist loads operating in a range of directions.
c) *Drainage*   The rain which falls on the roof is unwanted water and should be drained from the roof in the simplest and most direct manner. The roof form and coverings should be designed to achieve this.
d) *Durability*   The roof, being the highest part of a building, is generally the most inconvenient to gain access to, and the coverings should not, therefore, be affected by moisture, frost, atmospheric pollution, and other harmful agents which would cause a failure of the weather-exclusion properties or lead to heavy costs of maintenance. The effects of normal movement

should be catered for in the design, since solar changes affect the roof to a greater extent than other parts of the building.

e) *Thermal insulation*   Warm air, being less dense than cold air, rises; therefore in a heated building the greatest loss of heat will occur through the roof, unless suitable precautions are taken. The Building Regulations give a maximum coefficient of thermal transmittance for domestic roof construction (U-value = 0.6 W/m^2 $^\circ$C) so that energy used for heating is not wasted by excessive heat losses from that part of the building to the atmosphere.

f) *Sound insulation*   With the exception of special buildings such as concert halls, or buildings situated close to high external noise levels, e.g. near airports or motorways, the roof is not required to provide any sound insulation.

g) *Fire resistance*   The resistance of a roof to fire largely depends on its proximity to other buildings, and the primary function in this instance is to prevent the spread of fire to and from those buildings via the roof. A secondary function is to prevent fire spreading from one part of the building to another by way of the roof. Adequate resistance to fire is also required in order to prevent collapse of the roof before the occupants of a building have reached safety.

h) *Lighting*   Where large floor areas are covered by one roof, the natural light provided by windows in the external walls may be inadequate, and the standard of lighting may be improved by the incorporation of roof lights.

j) *Ventilation*   A means of ventilation through the roof may be required, especially in roofs which cover industrial buildings. Naturally occurring air currents within a building, improved by roof ventilation, can provide an economic means of removing noxious fumes from a manufacturing process.

k) *Appearance*   The appearance of the roof should harmonise with the surroundings and depends upon the shape of the roof, together with the shape and colour of the coverings. The appearance may be governed by the requirements of the local planning authority.

10.2 Basic roof forms
Roof forms may be classified in three ways: shape, span, and structural-design principle.

Shape
a) **Flat**   A flat roof is one where the slope in any plane does not exceed 10° to the horizontal (fig. 10.1).

b) **Pitched**   A pitch roof is one where the slope in any plane exceeds 10° to the horizontal. There are several types of pitched roof (fig. 10.1).

   i) *Single or mono-pitch*   This type is sometimes called a lean-to when it abuts a wall which rises to a higher level than the roof.

   ii) *Double-pitch*   This type is one where the roof slopes outwards in two directions.

   iii) *Butterfly*   In this case the roof slopes inwards in two directions.

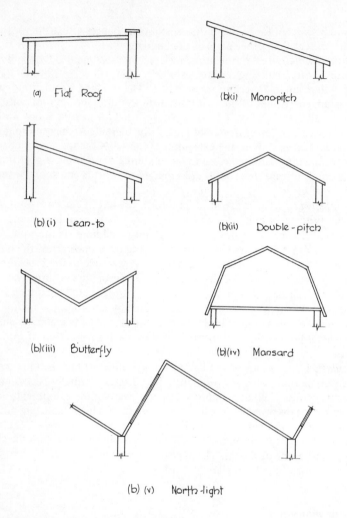

Fig. 10.1 Roof shapes (flat and pitched)

iv) *Mansard* Similar to the double-pitch, but the roof slopes have various pitches.

v) *North-light* A double-pitch roof in which the steeper slope is glazed and faces north, so that the light in the building produces little or no shadow.

c) **Curved** As the name implies, the roof surface is curved in one or more directions. There are again several types (fig. 10.2).

i) *Barrel vault* In this case the roof has the same curve in only one direction for its whole length.

Curved butterfly (gull-wing) roof

North-light folded-plate roof

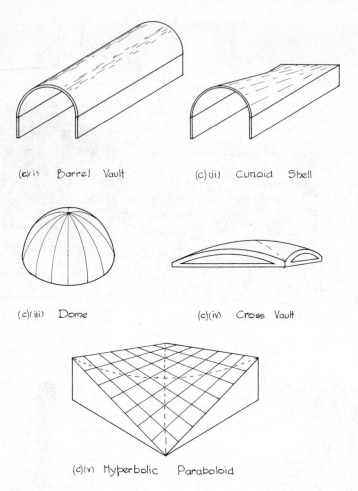

(c)(i) Barrel Vault   (c)(ii) Cunoid Shell

(c)(iii) Dome   (c)(iv) Cross Vault

(c)(v) Hyperbolic Paraboloid

Fig. 10.2 Roof shapes (curved)

ii) *Cunoid shell*   The roof has a reducing curve in one direction, the curve reducing over the whole length.

iii) *Dome*   This roof is in the shape of part of a sphere and rests on a circular base.

iv) *Cross vault*   The roof comprises two cylindrical shells which intersect to cover a rectangle.

v) *Hyperbolic paraboloid*   This is created by 'lifting' two diagonally opposite corners of a square to a higher level than the other two corners, and the name is derived from the curve in the roof surface which is produced between the diagonal corners — between the low corners a parabola is produced, and between the high corners a hyperbola.

122

North-light barrel-vault roof

Span

The span of a roof will have a major bearing on its shape, and can be classified in both distance and direction (fig. 10.3).

a) **Distance between supports**
   i) *Short span*   Up to 7.5 m.
   ii) *Medium span*   From 7.5 m to 24.0 m.
   iii) *Long span*   Over 24.0 m.
b) **Direction of span**   This applies equally to all shapes of roof and refers to the whole roof, not just one section.
   i) *Single roof*   The roof, or the main framework of the roof, spans in one direction only.
   ii) *Double roof*   The roof framework spans in two directions.
   iii) *Triple roof*   The roof framework spans in three directions.

Structural form

The method of supporting the roof coverings is generally by some form of framework (figs 10.4 to 10.6).

a) **Couple roof**   The simplest form of frame is that of the couple roof, where two inclined members rest against each other at their apex and are supported on walls at their lower ends.
b) **Close couple roof**   The problem with a couple roof is that the framing members or rafters tend to push outwards the walls on which they rest, and in order to restrict this movement the rafters are anchored at their lower end by another member known as a tie.

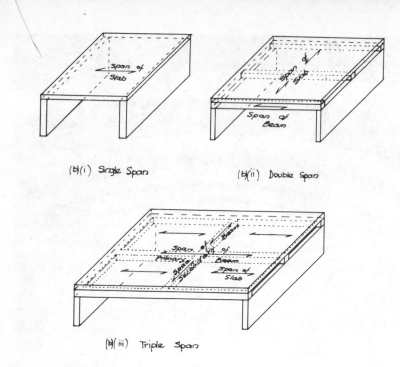

(b)(i) Single Span

(b)(ii) Double Span

(b)(iii) Triple Span

Fig. 10.3 Roof spans

c) **Collar roof** The rafters are still tied together to restrain them from open-ing, but the tie is situated near to the apex, thereby allowing more height to the room underneath, or alternatively a reduction in wall height while maintaining a suitable ceiling height for most of the room.

d) **Cruck roof** This is an example of the collar principle which was in use in the fifth century. Two tree branches were tied together to form an inverted 'V', a horizontal tie beam was then added to the crucks and was allowed to project out above the feet of the crucks, and vertical and longitudinal members were then added.

e) **Purlin roof** The longer the rafters become, the more likely they are to sag under the weight of the roof coverings which they carry. To assist in carrying the loads, longitudinal beams called purlins are introduced.

f) **Trussed rafter roof** The strength of the roof may be increased, the size of the members reduced, or the span of the roof increased if a framework is built up with a series of triangles.

The triangle is the only geometrical figure which cannot be distorted without changing the length of at least one of the sides.

g) **Rigid or portal frames** These are generally used for single-storey build-ings where a long clear floor area is required. In this case the roof frame is not supported on walls at high level, but is rigidly attached to its own

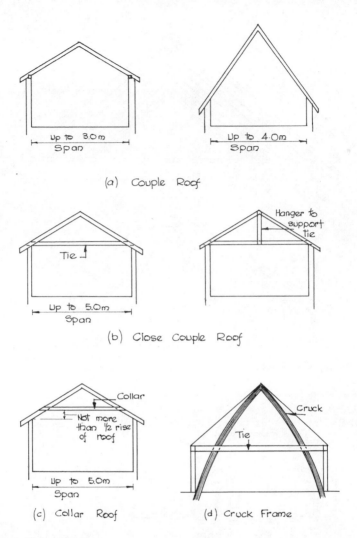

(a) Couple Roof

(b) Close Couple Roof

(c) Collar Roof

(d) Cruck Frame

Fig. 10.4 Structural forms

column support, thereby providing a completely rigid frame from the foundation.
h) **Folded plate or slab**   The strength of the roof is derived from its shape: a flat sheet will bend a considerable amount due to its own self weight, but the same plate folded will support not only its own weight but also additional weight. This principle has been applied for many years in corrugated sheets, but more recently the principle has been used for whole roofs.

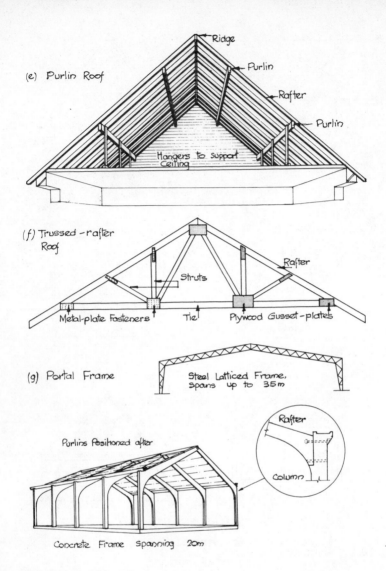

(e) Purlin Roof

Ridge
Purlin
Rafter
Purlin
Hangers to support Ceiling

(f) Trussed-rafter Roof

Struts
Rafter
Metal-plate Fasteners
Tie
Plywood Gusset-plates

(g) Portal Frame

Steel Latticed Frame, Spans up to 35m

Purlins Positioned after

Rafter
Column

Concrete Frame Spanning 20m

Fig. 10.5 Structural forms

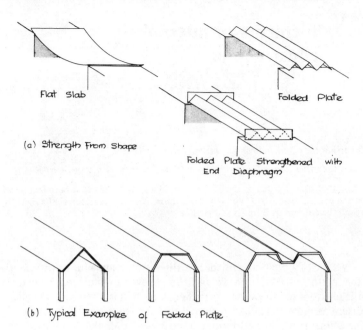

(a) Strength From Shape

Flat Slab

Folded Plate

Folded Plate Strengthened with End Diaphragm

(b) Typical Examples of Folded Plate

Fig. 10.6 Structural forms

11   Elements of internal construction

Identifies the elemental parts of internal construction.

11.1   Identifies internal walls, floors and stairs as primary internal elements.

Acknowledgement is due to the Technician Education Council for permission to use the content of the TEC units in this chapter. The council reserves the right to amend the content of its units at any time.

The envelope of a building provides weather protection and security, but without the internal construction it is of little use even for storage, since an earth floor is far from suitable. The internal construction is, therefore, equally as important as those elements which make up the envelope of a building.

The elements which are of prime importance inside the building are those which the builder erects along with the envelope or immediately afterwards, i.e. the floors, the stairs, and other walls. The ceiling is also classified under the SfB system as a primary element when it is a separate structure hanging down below a floor – this is known as a suspended ceiling.

The floors provide a suitable surface on which pedestrian and vehicular traffic can operate.

The provision of usable space on any given ground-plan area can be increased by the use of further floors set at suitable heights above and below the ground floor. This necessitates the movement of people in a vertical direction. Although lifts and ramps may enable this, it is more usual to provide stairs.

The normal method of partitioning a floor area into smaller units, where different operations or activities are carried out, is by means of further walls, known as internal walls. In certain large offices where noise levels are low, the subdivision is sometimes achieved by the judicious positioning of filing cabinets, pot plants, and other movable screens.

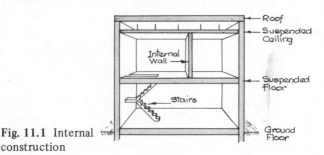

Fig. 11.1 Internal construction

12   Internal walls

Understands the functions of internal walls and the fundamentals of their construction.

12.1   States primary functions of internal walls.
12.2   Identifies essential differences between load-bearing and non-load-bearing walls.
12.3   Sketches and describes typical brick and block internal walls including details around openings.
12.4   Describes various finishes for internal walls.

Acknowledgement is due to the Technician Education Council for permission to use the content of the TEC units in this chapter. The council reserves the right to amend the content of its units at any time.

12.1 Functions of internal walls

The primary functions of an internal wall are

a) to assist in carrying the loads imposed on the structure,
b) to partition space,
c) to provide insulation, both sound and thermal,
d) to provide fire resistance,
e) to provide security,
f) to support fixtures, fittings, and decorations,
g) to accommodate services,
h) to be static or movable.

a) The design of the structure may require some of the floor, roof, and wind loads to be carried by walls within the envelope, as well as by those walls which form the envelope. In such a case, the criteria for design are similar to those of the external wall, i.e. mainly strength and stability.
b) The wall may be provided to enclose space, i.e. to form rooms or to subdivide rooms into smaller areas.
c) In many cases, the ownership of a building is in the hands of one person or company, in which case, with the exception of specialist operations such as cold storage, recording studios, etc., there is little need for high levels of insulation between different areas of the building. There are instances, however, as in apartment buildings and blocks of flats, where it would be beneficial to the parties on either side of the common wall to have both thermal and sound insulation.
d) The containment of fire is of prime importance in reducing the effect on the structure and its contents, therefore any walls which form rooms or

corridors should provide adequate fire resistance – sufficient to allow personnel to escape and the fire brigade to gain access without incurring personal hazard.

e) Security is essential in multiple-occupancy buildings but not necessarily of importance in the traditional house, where the external walls perform that function.

f) In many office and domestic situations, cupboards, shelves, and other storage units are hung and fixed to the walls (fig. 12.1), therefore the wall may have excessive loads attached to it which must be catered for, both in load-bearing and in fixing terms. The surface of the wall should also be suitable to receive a finish, if required.

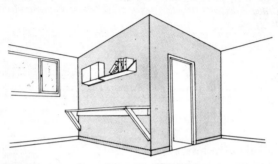

Fig. 12.1 Partitioning

g) In the majority of buildings, services such as electric lighting and power points, gas outlets, water pipes, telephone and television outlets, and central-heating ducts or pipes must pass through, be let into, or be fixed to the walls. The material of the wall must therefore be able to accommodate, without serious loss of strength, holes being drilled through it, chases cut in it, or screws and nails driven into it.

h) The layout planning of buildings is a major consideration, and, in order to achieve maximum usage of the floor space, it is frequently desirable to be able to move, remove, or reposition internal walls to suit various requirements. It is obvious that in these cases the walls should not be carrying any structural loads, since the removal of such walls would create a large amount of work and necessitate additional structural design.

12.2 Load-bearing and non-load-bearing walls (fig. 12.2)

A wall which carries some structural load is known as a load-bearing wall, and it should have a suitable foundation, on either a concrete strip or raft. The material must have sufficient strength and thickness to provide support and stability for the loads; hence such walls are commonly constructed of bricks, blocks, or concrete.

Those walls which are designed to enclose space and not support any structural loads are known as non-load-bearing walls, or partitions, and are usually

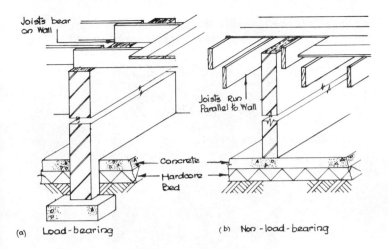

Fig. 12.2 Internal walls

only one storey in height. They may be constructed of bricks or blocks, but, in order to reduce costs or provide the mobility previously mentioned, other forms of lighter-weight construction are used.

12.3 Brick and block internal walls, and details

Internal walls of brick and block are bonded for strength, as mentioned in Chapter 8. The walls are of $\frac{1}{2}$B (102.5 mm) or 1B (215 mm) thickness, as there is generally a lighter loading on them than on the external walls, and they are solid since there is no need for weather resistance. In modern domestic two-storey construction, the majority of these walls are only one-storey high, as they provide support for the first floor but not for the roof structure. The bonding of the wall is usually in stretcher bond, and this continues both at the corners and at other junctions (fig. 12.3).

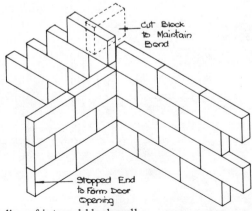

Fig. 12.3 Bonding of internal block wall

131

Block wall, showing bonding at junction and stopped ends

Where an opening in an internal wall is required for a doorway or a serving hatch, the opening must be of a suitable size to accommodate the component without a lot of additional work being incurred. In many instances the builder will incorporate the door frame in the wall as the work proceeds (fig. 12.4), but this is possible only where the frame has been prefabricated. It is a better

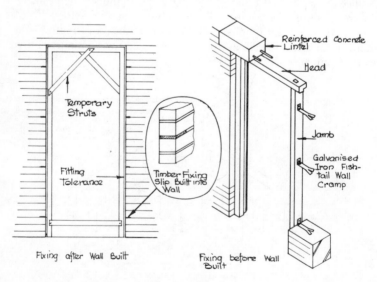

Fig. 12.4 Details at door openings

Building in door frames

method of construction because the frame can be solidly anchored into the wall by means of fixing cramps rather than by the poorer alternative of screws and nails. There are no fitting tolerances which require making good, since the wall is built up to the already positioned frame. The frame is held in a rectangular shape by temporary thin softwood struts tacked across the angles.

The height of the frame, unless it is a storey-height frame (see fig. 9.14), is less than that of the floor-to-ceiling dimension, This means that the wall must be continued above the frame and, as the framing is not designed to carry the weight of walling above, some additional support must be provided. This is supplied by a lintel or by reinforcing the brickwork immediately above the frame with wire mesh – there are therefore many sizes and shapes of lintel, depending upon the load to be supported over the opening, the architectural preference for materials and manufacture, and the nature of the wall finish:

a) *Precast concrete*   Heavy to lift into position, especially when the span is large, but having adequate strength.

b) *In-situ concrete*   A 'long' process which holds up the laying of bricks or blocks because of the time taken to fix and pour the concrete and the length of time required for the concrete to harden and strengthen.

c) *Timber*   Suitable for most openings other than large spans, easily fixed in position, and easy to fix other finishes to.

d) *Steel*   Strong, lightweight, will not rust if galvanised, and easily fixed. Requires special treatments or preparations in order to receive certain finishes.

133

12.4 Finishes for internal walls

There are a variety of finishes for internal walls and their selection will depend upon the function of the building, the decorative effect required, and the cost (fig. 12.5).

The most common internal wall finish is plaster work, which provides a smooth dense jointless surface suitable for painting or decorating. Plaster also increases the resistance of a wall to fire and flame-spread. Plaster work is dealt with more fully in volume 2 of this book.

Plasterboard sheets 2.4 m x 1.2 m x 12.7 mm are used as a finish to brick and block walls; this is a 'dry' alternative to the 'wet' operation of plastering. Softwood timber battens 38 mm x 19 mm are fixed vertically to the wall at 400 mm centres, using screws or nails, and the plasterboard is fixed to the battens using galvanised steel nails. The joints between boards are reinforced with paper reinforcing tape or scrim and the joint is filled flush with the surface of the board, or the whole wall area is finished with a 3.5 mm skim coat of plaster.

Wallboards, plywood, fibreboard, flat asbestos-cement sheet, and thin sheets of plywood or hardboard which have a surface facing of laminated plastics glued to them are fixed to battens by panel pins, nails, or screws, and the battens are positioned and fixed to the wall in exactly the same way as for plasterboard. The joints between boards are formed dry and are either featured or concealed.

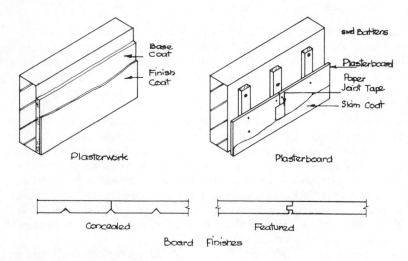

Fig. 12.5 Internal-wall finishes

Wall tiles are manufactured in a variety of materials, the most common being ceramics, glass, plastics, and cork, although other tiles usually associated with floor finishes may be used. The tiles, while improving the decorative effect, may also be used for other purposes such as water resistance in bath-rooms and kitchens, abrasion resistance in hospital corridors, and sound in-sulation in typing-pool offices. For further details see volume 2.

The brick or block wall may be left uncovered, in which case the mortar joints may be featured by using coloured mortars, by using special pointing (see fig. 8.14), or by a combination of both. Special lighting can further en-hance the decorative appearance of such walls.

As construction costs increase, the quality of finishings in functional build-ings is reduced and may be dispensed with altogether in order to save unneces-sary expense. Walls left exposed and having no requirement for decorative effect have their mortar joints flush-jointed and finished off by being rubbed over with hessian or paper bags — known as 'bagging off'. This forms a smooth servicable surface which is suitable for painting if required at some time in the future.

Paints may be applied to the bare walls or to other previously mentioned finishes such as plaster, plasterboard, and wallboards. They are also described in more detail in volume 2.

13   Floors

Understands the functions of floors and the fundamentals of their construction.

13.1   States primary functions of ground floors.
13.2   Sketches and describes the junction between a solid ground floor and external wall.
13.3   Identifies the need for hardcore and names common materials in use.
13.4   Sketches and describes the junction between a suspended timber ground floor and external wall.
13.5   Discusses the relative advantages and disadvantages of solid and suspended ground floor.
13.6   Sketches and describes the construction of a suspended timber upper floor including trimming for openings.
13.7   Identifies the need for strutting suspended timber upper floors and sketches typical details.
13.8   Identifies appropriate floor finishes for domestic construction.

Acknowledgement is due to the Technician Education Council for permission to use the content of the TEC units in this chapter. The council reserves the right to amend the content of its units at any time.

13.1 Functions of a ground floor
The primary functions of a ground floor are
a) to carry the loads which may be imposed on it,
b) to prevent dampness entering the building from the ground,
c) to provide thermal insulation,
d) to prevent the growth of vegetable matter within the building,
e) to provide a suitable wearing surface.

a) The floor carries the majority of the internally imposed loading and must therefore have sufficient strength and stability safely to carry and transmit those loads to the walls or the subsoil. The loads which are imposed vary with the nature of occupancy of any given room, and the floor should be designed accordingly. British Standard Code of Practice CP3:Chapter V: Part 1 sets out in tabular form the recommended design loading for floors in a wide range of uses. Additional precautions must be taken where excessively heavy point loads may be applied to the floor in the form of machine beds or load-bearing columns or walls.
b) The ground-floor construction is in many cases in contact with the subsoil, and dampness will rise from the ground by capillary action, entering the

building from underneath. The construction must therefore be such that this passage of moisture is prevented. Building Regulations clause C3(i) states 'such part of a building as is next to the ground shall have a floor so constructed as to prevent the passage of moisture from the ground to the upper surface of the floor.'

c) Although warm air rises, heat can nevertheless be conducted from the building by floors in contact with the ground. It is therefore preferable, although not essential, to incorporate some form of insulation in the floor structure, especially where underfloor heating is to be installed.

d) In damp and humid conditions, the growth of vegetable matter flourishes. It was stated in Chapter 6 that topsoil is removed from the site before the start of construction work, but this does not mean that all roots will have been removed, nor does it mean that seeds will not find their way on to the cleared site. It is therefore essential to prevent the growth of vegetable matter which may in time have a detrimental affect on the materials used in floor construction. There are three methods of preventing growth:
   i) removing the humid conditions by suitable ventilation,
   ii) preventing growth by covering the area with a dense concrete layer 100 mm thick,
   iii) a combination of (i) and (ii).

e) The suitability of a wearing surface, the surface which initially carries the load, will depend upon the use to which a particular floor will be put, i.e. wearing surfaces which are suitable for a living room may not be suitable for a factory, because the natures of the loading and use are different.

13.2 Solid ground floors

There are two types of ground-floor construction: solid and suspended. A solid ground floor is one in which the whole of the floor area is in contact with the subsoil. A suspended floor is one which spans over a certain area.

A solid ground floor comprises several layers of material which are built up from the formation level, the layers being hardcore, blinding, damp-proof membranes, concrete slab, wearing surface, and finish (fig. 13.1).

Fig. 13.1 Solid-floor construction

Hardcore is spread, levelled, and rolled over the formation to provide a suitable base for the construction. The hardcore comprises large particles which have rough edges, and the surface, even after rolling, is unsuitable to receive in-situ concrete. The main reason is that when the concrete, especially that having a high water:cement ratio, is compacted, the water together with the fine particles of cement and sand (known as grout) will 'flow' down into the voids in the hardcore, thereby reducing the strength of the concrete. It is therefore necessary to prevent this loss of grout by sealing the voids or by introducing some form of barrier membrane. However, the membrane may well be punctured during the concreting operation if it is laid directly on the jagged hardcore. The blinding is either fine quarry waste or dry weak concrete which is placed over the hardcore, serving both to fill the voids in the hardcore in the surface region and also to provide a relatively smooth surface on which to cast the slab.

The damp-proof membrane (d.p.m.) may be placed either above or below the slab. If placed below, the d.p.m. is usually in the form of a polythene sheet which will not only act as a d.p.m. but will also prevent loss of grout from the concrete slab when cast. If placed on top of the slab, the d.p.m. may be in the form of a polythene sheet, a coating of bituminous paint, or asphalt. In all cases of solid ground-floor construction there is a possibility of moisture in the ground bypassing the damp-proof course (d.p.c.) in the foundation walls, and it is therefore essential that the d.p.m. in the solid ground floor is lapped with the d.p.c. in the surrounding walls (fig. 13.2).

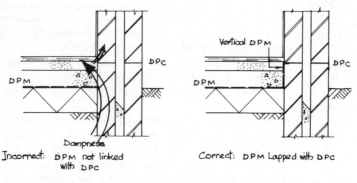

Fig. 13.2 Prevention of dampness through floor

The concrete slab is cast to a depth suitable for transferring the floor loads to the ground, the surface being finished in a variety of ways depending upon the type of construction to be placed on top. A tamped finish provides a rough surface to which the wearing surface can easily be keyed, while a floated finish provides a smooth surface to which a surface d.p.m. can be applied. The slab acts as a raft, in a similar manner to a raft foundation, and should not be built into the surrounding walls, because, if any ground settlement took place under the slab, a void would be formed and the slab would be subjected to loading conditions for which it had not been designed.

The wearing surface usually takes the form of a cement–sand screed, 1:3 mix, the thickness of which varies from 25 to 65 mm. This screed serves to
a) take out any surface irregularities in the concrete bed,
b) provide a suitable surface to receive the finishes,

c) adjust the level of the floor so that differing floor-finish thicknesses are accommodated,
d) provide insulation,
e) accommodate service pipes and cables.

13.3 Hardcore

The purpose of hardcore is
a) to ensure a consistent material over the whole floor area so that the loading of the floor is uniformly spread over the whole area;
b) to reduce the capillary action of moisture from the ground because of the size of voids within the layer;
c) to fill areas which, after topsoil removal and reduced-level excavation, are below the level required for floor construction;
d) to provide a clean, dry, and firm working surface.

Where a large amount of fill is required, the hardcore should be laid in layers approximately 150 mm thick and be well consolidated to prevent future settlement.

In order to prevent movement within the hardcore, the material used should not be soft or easily crushed, nor,.in accordance with Building Regulations clause C3(3), should it be affected by the pressure of ground water or sulphates. The hardcore should contain little or no fine material and should be inert. The materials commonly used for hardcore are as follows.

a) *Ceramic rubble*　Bricks and tiles taken from buildings which have been demolished; but care should be taken to ensure that there is no plaster or timber in the rubble, as these materials will be affected by the moisture in the ground.
b) *Clinker*　This is sintered or fused ash from furnaces.
c) *Pulverised fuel ash* (p.f.a.)　Produced from the pulverised coal used at power stations, it is a lightweight material which, when mixed with a specific amount of water and compacted, acts like a weak cement.
d) *Gravel*　Crushed rocks, ideally well graded, which make excellent but expensive hardcore.
e) *Quarry waste*　Cheaper than gravel since it is not uniformly graded, it is nevertheless clean and hard, but requires more consolidation than gravel.
f) *Shale*　A naturally occurring material created by the baking of clay, millions of years ago. Care must be taken in the selection of shale for use as hardcore, since certain types are very weak.

13.4 Suspended ground floors

A suspended ground floor is one which is not directly in contact with the ground but spans from one wall to another. The walls which support the floor may continue up to higher levels or may be constructed to support only the ground floor — in the latter case, these walls are known as dwarf or sleeper walls.

After removal of the vegetable soil from the site and construction of the external load-bearing walls, a layer of hardcore is spread, consolidated, and topped with suitable blinding material over the floor area. A bed of concrete is then placed on the hardcore and the top surface is smoothed over with the back of a spade or shovel. In accordance with the Building Regulations, the concrete should consist of one part cement to three parts fine aggregate to six parts coarse aggregate by volume, be at least 100 mm thick, and have its top surface at or above the highest level of the surrounding ground.

The sleeper walls may have their own strip foundations or may be constructed from the concrete bed. The regulations require that there is a space above the upper surface of the concrete of not less than 75 mm to the underside of the suspended timbers, and such space must be clear of debris and have adequate through ventilation (fig. 13.3).

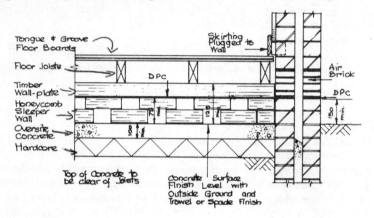

Fig. 13.3 Suspended-floor regulations

In order to provide adequate ventilation, air-bricks must be inserted in the external walls of the building, and the sleeper walls are constructed in a honeycomb manner, thereby allowing free air circulation.

Ventilation is essential in order to keep the moisture content of the timber floor members below 20% of their dry weight. Above that level, fungal growth may take place, causing a weakening of the timber and leading to eventual collapse.

A damp-proof course is placed on top of the sleeper wall so that moisture from the ground cannot reach and affect the timber wall-plate which spreads the load from the floor to the wall.

The floor consists of timber boards or other suitable sheet material such as chipboard, supported by bearers or joists (fig. 13.4). These joists are spaced at 400–600 mm centres, the spacing depending on the weight which the floor has to support, the size of the joists, and the span of the joist from one wall to the next. A table in the Building Regulations indicates suitable dimensions for various loading conditions.

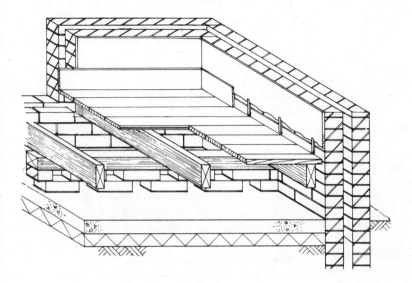

Fig. 13.4 Typical suspended ground floor

The boards should be rift-sawn, which means that they have been cut radially from the log, thus reducing the amount of movement which may occur in the board as a result of change in temperature or moisture content. This, however, is an expensive method of converting timber and it is more usual to use tangential-sawn boards (fig. 13.5). These are more likely to warp, which results in greater wear.

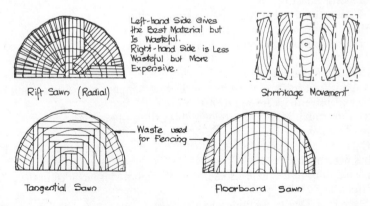

Fig. 13.5 Timber conversion

The majority of softwood boards are fixed at right angles to the joists and have tongued-and-grooved joints. They should be cramped tightly together

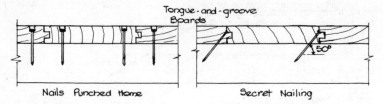

Fig. 13.6 Floorboard fixing

and then nailed to the joists, the nails being punched well below the surface (fig. 13.6).

It is usual to leave a small gap between the timber floor and the surrounding walls; this allows the floor to expand, contract, and flex without rubbing against the wall, and at the same time prevents direct contact of the timber with a surface which may be damp. The gap is masked by the skirting board which is fixed to the wall.

13.5 Relative merits of solid and suspended ground floors

Solid floors – advantages
a) The floor is completely damp-proof, provided the membrane is not punctured.
b) The floor and raft foundation can be the same unit.
c) There are no under-floor draughts.
d) There is no likelihood of rot occurring in the construction.
e) No expansion or contraction joints are required because of slab size in domestic construction.
f) The construction has good fire-resisting qualities.
g) Heat losses can be reduced to acceptable levels.
h) It is a cheap form of construction on level sites.

Solid floors – disadvantages
a) The slightest break in the d.p.m. will allow dampness to penetrate and will be expensive to locate and rectify.
b) The construction is expensive on sloping sites, where a large amount of fill would be required.
c) Once completed, services are difficult to alter or increase, unless provision has been made.
d) A cold floor results unless some form of insulation is included in the construction.
e) There is no facility for under-floor storage.
f) The floor, being dense, will transmit certain sounds.

Suspended floors – advantages
a) The floor is unlikely to be affected by dampness.
b) Where any deterioration takes place, it can soon be treated and remedied.

c) Services are easy to locate and adapt.
d) The floor can flex and take up small ground movements.
e) The floor will readily accept other fixings.
f) The construction will be cheaper on sloping sites.
g) There is a facility for under-floor storage.

Suspended floors – disadvantages
a) Deterioration will take place if either the ventilation is cut off or excessive dampness is present.
b) Insulation is required to overcome heat losses resulting from under-floor draughts.
c) The construction does not have good-fire resisting qualities.
d) The construction is expensive on level sites.

13.6 Suspended timber upper floors

The construction of a suspended timber upper floor is similar to that of the ground floor in the use of joists and boards; however, there are no honeycomb sleeper walls, the joists being built into or on to the load-bearing internal and external walls. There is also a need to travel from one level to another, and the floor must be constructed in such a manner as to accommodate stairs, lifts, flues, hearths, etc. This form of construction is known as 'trimming to openings' (fig. 13.7).

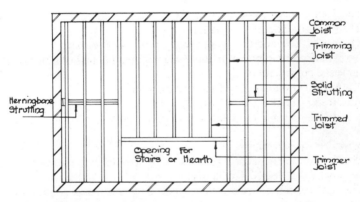

Fig. 13.7 Trimmings to openings

The sizes of joists used for the upper-floor construction are larger than those used for the ground floor, because the spans have increased, and those members which form the framework around an opening must be even larger as they support the extra load in that area. In order to maintain a level floor and soffit, this increase in size must be an increase in width, usually of 25 mm. To keep costs within reason wherever practicable, the joists should span the shortest distance.

143

The joists which are not affected by the opening are known as the common joists, and those which have been shortened are known as the trimmed joists. The member of the framework surrounding the opening which supports the trimmed joists is known as the trimmer joist, and the joists which complete the framework and support the trimmer joists are known as the trimming joists.

There are several methods of jointing the various joists around the opening (fig. 13.8). The tusk tenon joint is used to joint the trimmer joists to the trimming joint. It is a strong joint, so designed that cutting away of timber occurs in the middle section of the trimming joist, where there is least stress. This joint requires a lot of work in its formation, and is therefore expensive. More common joints are the housed joint, the bevelled housed joint, the ship-lap joint, or the joist hanger.

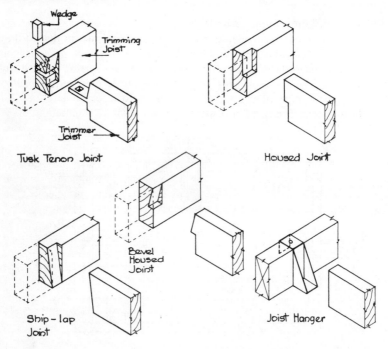

Fig. 13.8 Timber-joist connections

Joists which run parallel to and against a wall should be kept approximately 50 mm away from the wall — this allows for a proper fixing of the finishes both above and below the joist, as well as preventing the possible transmission of any moisture from the wall to the joist.

There are many methods of supporting the joists at a load-bearing wall, and the two current methods most commonly used are those of building the joist

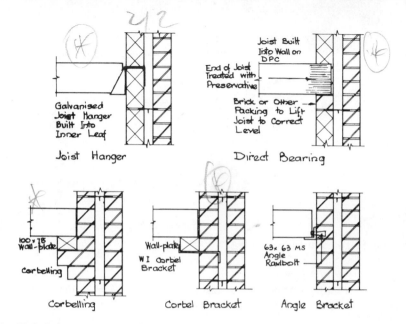

Fig. 13.9 Joist-support details

into the wall or supporting the end by means of a joist hanger (fig. 13.9). In the first instance, the ends of the joists should be treated with preservative to prevent deterioration of the timber by the presence of both moisture in the wall and, in the case of cavity-wall construction, dampness in the cavity. The joists should be packed up to level by means of slate or other damp-proof-course material.

Galvanised mild-steel joist hangers can provide an economical alternative method of support by reducing the length of joist required, and, by keeping the joist away from the wall, the level of the joist may be adjusted by inserting a packing in the shoe of the hanger or by adjusting the bend section which is inserted into the wall.

Other methods of supporting the floor joists include corbelling the brickwork or using a wrought-iron corbel to support a wall-plate, or a piece of mild-steel angle bolted to the wall.

13.7 Strutting

As the span of a joist increases from 3.0 m up to a maximum economical distance of approximately 4.5 m, there is a tendency for the joists to twist or buckle under load, which reduces their strength. In order to prevent this twisting action and to stiffen the construction, strutting is used.

There are two methods of strutting: herring-bone strutting and solid strutting (fig. 13.10).

a) *Herring-bone strutting*   This consists of timbers approximately 40 mm x 50 mm in cross-section, having their ends cut on a splay, fixed by nails

145

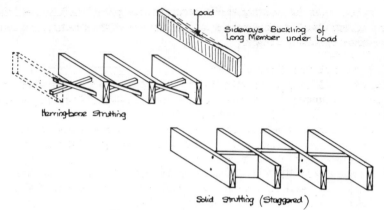

Fig. 13.10 Strutting of joists

between the top edge of one joist and the bottom edge of the next and also nailed to each other where they intersect.

b) *Solid strutting*   This consists of timber approximately 32 mm thick fixed between the joists, and the struts are either staggered to allow for end nailing or are in line, in which case skew nailing is used. This method is not as good as the herring-bone, but has the advantage of being easier and cheaper to construct.

Suspended-floor construction with herring-bone strutting

In both cases the strutting usually occurs at mid-span of the joists, and the gap between the last joists and the wall, on the line of the strutting, is secured by means of folding wedges.

13.8 Domestic floor finishes

There is a wide range of materials which are used as coverings or finishes; some are more commonly found in industrial or commercial buildings, but this does not preclude their use in a domestic situation.

Before a decision is reached as to the most suitable covering to be used in a given situation, it is necessary to consider what is required of that finish. The main considerations are as follows.

a) *Resistance to wear*   This includes the resistance to indentation caused by heavy furniture or stiletto heels, and resistance to abrasion caused by dirt being ground into the surface by shoes or wheels.

b) *Resistance to oil, grease, acids, and alkalis*   This is particularly important in kitchens and store-rooms.

c) *Resistance to surface moisture*   This is important in bathrooms, entrance halls, and kitchens.

d) *Resistance to stains and dirt*   This is again important in entrances and dining areas, and can also be linked with item (e).

e) *Ease of cleaning*   The modern housewife does not want finishes which take a long time to clean, or which require expensive cleaning equipment or chemicals.

f) *Resilience*   A hard surface is not generally suitable for domestic situations.

g) *Appearance and warmth*   Some finishes appear much warmer than others, and colours can also play an important part in creating a particular feeling.

h) *Durability*   The finish should be durable and require little maintenance.

j) *Cost*   The cost generally depends on the quality of the finish. High initial costs usually go hand in hand with long life and low maintenance costs, while low initial costs indicate short life and high maintenance charges. The maintenance includes costs of repairs, cleaning, and retreating surfaces.

Floor finishes can be classified into four groups: stone, wood, sheet and tile, and carpet. The suitability of a particular finish depends upon the situation, i.e. position, use, and requirement.

Stone floor finishes (fig. 13.11)

This group may be further subdivided into natural and artificial finishes. The artificial are those finishes which are man-made, and include quarry tile, brick, concrete, terrazzo, and granolithic.

a) *Natural*   Generally laid in slabs of varying size and thickness, the stone is bedded in mortar or cement grout on a screeded concrete subfloor — the bedding plane of the stone should always be parallel to the floor. It is a suitable finish in areas of heavy foot traffic, such as entrance halls. The hardness of the stone depends upon type, and this varies from soft sandstone, through limestone and marble, to very hard granite.

b) *Brick and tile*   Tile sizes vary from 25 to 150 mm square and 12–25 mm thick, and a variety of surface finishes and colours are available. The finish

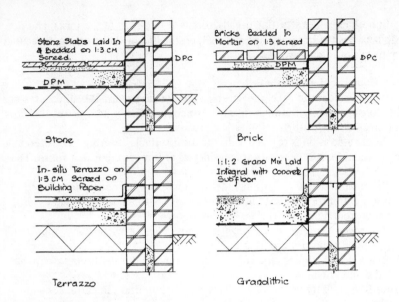

Fig. 13.11 Stone floor finishes

is laid on a screeded concrete subfloor and is jointed with mortar or cement grout. It is suitable for areas where water or chemicals are present, as in kitchens and toilets.

c) *Concrete*   A suitable finish in basements and garages, it should be laid in bays of approximately 50 m². The surface finish may be tamped, giving a rough surface for traction; wood-floated, providing a smooth non-slip surface; or steel-trowelled, which gives a smooth glossy finish.

d) *Terrazzo*   Comprising a carefully graded crushed marble embedded in a white or coloured cement matrix, it can be laid in situ or in tile form on a screeded concrete subfloor, and is suitable for wet areas and entrances.

e) *Granolithic*   Laid directly on a concrete subfloor, it is a fine concrete which uses a hard aggregate such as ground granite. While it is a more suitable finish for heavy-duty industrial floors, it can also be used in domestic entrances and porches.

Wood floor finishes (fig. 13.12)

This group may also be subdivided into natural and artificial finishes. The natural finishes comprise strip and board, block, parquet, and mosaic, while the artificial include cork and chipboard.

a) *Strip and board*   These floors are usually tongued and grooved and fixed to timber bearers or joists by means of nails. Softwood boards can be surface-nailed because they are usually covered by another finish such as carpet or linoleum, whereas hardwood boards are secret-nailed because no further covering is required (see fig. 13.6).

148

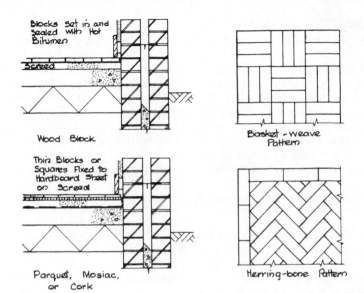

Fig. 13.12 Wood floor finishes

b) *Wood block*   These floors comprise blocks up to 90 mm wide, 150–300 mm long, and 25–38 mm thick, secured by mastic or bitumen emulsion mixed with rubber latex, and laid in patterns on a screeded floor.

c) *Parquet and mosaic*   Parquet sections of hardwood panels are between 300 and 600 mm long and 6–10 mm thick, and are laid in various designs on a smooth even floor, with glue or panel pins. Mosaic flooring is similar to parquet, but the hardwood blocks are of smaller size being about 100 mm long x 25 mm wide x 10 mm thick.

All natural wood floors are cheap, relatively easy to lay, resilient, and soft and quiet to walk on, and are therefore suitable for most domestic situations. It is essential, however, that the wood is properly seasoned and that the moisture content is suitable for the given position, otherwise shrinkage or swelling will occur. The table shown in fig. 13.13 indicates the range of moisture content for different situations.

Situation	*Moisture content* (%)
Framing and sheathing	23 – 21
External joinery	18 – 16
Internal joinery: intermittent heating	15
continuous heating	12 – 9
close to heating source	8
Floor finish above under-floor heating	8 – 6

Fig. 13.13 Table of moisture contents for timber

d) *Cork*   The bark from the cork or cork oak tree is ground into small
irregular granules and these are compressed and heated, thus releasing
natural resins which, when combined with synthetic resins, bond the cork
particles together. Laid on a dry screeded subfloor, the tiles, of various
sizes ranging from 100 to 900 mm square and 3–14 mm thick, are secured
with headless steel pins or adhesive. The finish is soft, decorative, and has
good thermal- and sound-insulation properties.

e) *Chipboard*   Composed of graded wood chips and plastic resins which are
subjected to heat and pressure to form boards, these boards are laid either
butt-jointed or with loose plastics tongues inserted in preformed grooves
around the edges. Care should be taken to see that the boards are adequately
supported and not subjected to moisture. This finish is more economical
than the natural strip floor in most domestic situations.

Sheet and tile floor finishes (fig. 13.14)

Included in this group are mastic asphalt, pitch mastic, linoleum, thermo-
plastics, p.v.c., and vinyl asbestos and rubber tiles.

a) *Mastic asphalt*   This material is a combination of bitumen, inert material
such as clay or silica, and graded mineral aggregates, heated to a temperature
between 190 and 220 °C and spread with wooden floats. It is generally laid
in one or two coats, the thickness depending on use (16 mm for domestic,
25–32 mm for industrial). The thinner coats should be laid on an isolating
layer of sheathing felt to avoid cracking which would result from differential movements of the subfloor and the finish. Mastic asphalt, when laid,
is impervious and may be used as a finish in wet areas such as washrooms
and sculleries. It can also be used as a d.p.m. on solid ground floors under a
thin sheet or tile finish such as p.v.c.

b) *Pitch mastic*   This comprises aggregates of graded limestone, other rock
and stone fragments, coal-tar pitch, flux oil, and lake asphalt. The lake
asphalt is added to make the pitch mastic more resilient to cracking and
indentation. The finish is laid in a similar manner to mastic asphalt, but has
a better resistance to oils, fats, and grease and is therefore suitable for use
in kitchens.

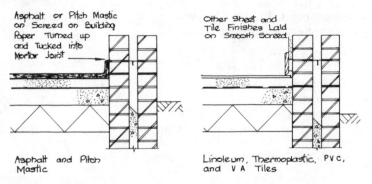

Fig. 13.14   Sheet and tile floor finishes

150

c) *Linoleum*   This material has a backing of jute canvas, hessian, felt, or bituminised paper felt on to which is pressed a topping composed of linseed oil, gums and resins, cork or wood flour, and pigments. A wide variety of patterns can be achieved by self-colouring, printing, or embossing. Linoleum is produced in sheet or tile form; the sheet width is usually 1.8 m, the tile size either 225 or 300 mm square, with the thickness varying from 2 to 7 mm. This material can be laid on most floor surfaces, using an adhesive, but if rising dampness is present a d.p.m. should be laid between the sub-floor and the linoleum. Although linoleum is quiet and resilient, it may be cut if subjected to point loads or sharp edges, but is suitable for most domestic situations.

d) *Thermoplastic tiles*   These are produced from asbestos fibres, asphaltic or synthetic resin binders, and pigments, mechanically mixed and formed under heat and pressure into tiles 225 mm square and 3 or 5 mm thick. These tiles are moisture-resistant and have good thermal-insulation and wearing properties. They are heated to make them flexible prior to laying on a smooth surface with a bituminous adhesive.

e) *P.V.C.*   This material is composed of polyvinyl chloride resin, plasticisers, stabilisers, and pigments, and it is produced as sheet (900–2400 mm wide) or tiles (300 mm square) in thicknesses ranging from 1.6 to 3 mm. The sheet p.v.c. can be welded at the seams to form a jointless floor, and the material may be unbacked or have a backing of felt, plastics foam, fabric, or cork. The sheet or tiles are laid on a smooth subfloor with a suitable adhesive, and provide a pleasing and servicable finish in most domestic situations where the temperature of the floor does not exceed 29 °C.

f) *Vinyl asbestos*   Similar to p.v.c. but with asbestos fibres added, this is produced in tiles either 225 or 300 mm square with thicknesses ranging from 1.6 to 3 mm. These tiles have a better grease-resistance than p.v.c. and are therefore better suited to use in kitchens.

g) *Rubber*   Produced from natural or synthetic rubber with mineral fillers, such as china clay, pigments, and anti-oxidants, as both sheet (900–1300 mm wide) and tiles (150–600 mm square). Thickness ranges from 3 to 13 mm, with 5 mm being the most common. The surface finish can be plain, ribbed, ridged, or studded, and in some instances the under-side is also ribbed or ridged to improve adhesion. Sponge-rubber flooring comprises a solid rubber wearing surface with sponge or foam backing – occasionally fabric is placed between the layers. The material is laid on a smooth surface using a rubber-based adhesive; the finish is soft, quiet, water- and bacteria-resistant, and suitable for nurseries, bathrooms, kitchens, and corridors.

Carpets (fig. 13.15)
There are many types of carpets available to suit most situations, and they can be classified by their use, by the materials used, or by their method of manufacture.

a) *Use or purpose*   Carpets are graded into five categories depending on the anticipated wear.

Category	Area of building
i) Light domestic	Bedrooms
ii) Medium domestic	Dining and living rooms, hotel bedrooms, waiting rooms
iii) Light contract and general domestic	Private staircases, halls, offices
iv) Medium contract and heavy domestic	Classrooms, canteens, corridors
v) Heavy contract	Main corridors, lobbies, high-circulation areas

b) *Materials*   The fibres used in manufacturing carpets can be classified into four categories.
  i) Natural vegetable fibres — cotton, sisal, jute
  ii) Natural animal fibres — wool
  iii) Synthetic fibres — nylon, acrylic, polyester, polypropylene, viscose rayon
  iv) Mineral fibres — Stainless steel, aluminium

c) *Manufacture*   Carpets are manufactured on machinery which does not limit the length, the width varying from 685 to 5485 mm. Carpet tiles are also available in 400, 500, and 600 mm square sizes. The common methods of manufacture are (i) woven, (ii) tufted, (iii) needle-pinched, (iv) flocked, (v) fibre-bonded, (vi) knitted.

   Carpets should be laid in dry situations as their moisture resistance is poor, and in order to increase the life of a carpet an underlay or resilient backing should be used. Underlays also increase the insulation properties of the covering and consist of felt, rubber, plastics foam, or rubberised felt. Carpets are secured to the floor by nailing around the perimeter, by adhesives, or by stretching over fixing strips or anchorages previously fixed to the floor. Carpet tiles may be fixed, but are usually laid loose, relying on the fibres around the perimeter of the square interlocking with those of the surrounding squares.

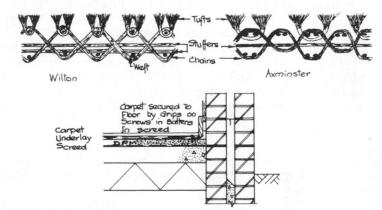

Fig. 13.15 Carpets

152

14   Stairs

Understands the functions of stairs and the fundamentals of their construction.

14.1   States the functions of a stair.
14.2   Identifies the constituent parts of a straight flight stair.
14.3   Identifies rise, going, handrail height and headroom as critical dimensions of stair construction.
14.4   Sketches and describes the construction of straight-flight stairs.

Acknowledgement is due to the Technician Education Council for permission to use the content of the TEC units in this chapter. The council reserves the right to amend the content of its units at any time.

14.1 The function of a stair
The function of a stair is to allow easy pedestrian access and egress from one floor level to another in safety. The ease of movement is dealt with later on in the chapter. The safety aspects are two-fold; firstly, the stairs must be strong enough to carry not only the weight of people using them but also the weight of any furniture and equipment being moved from one level to another; secondly, the stair must withstand the effects of fire for a sufficient length of time to provide a safe means of escape for the occupants of a building.

14.2 Constituents of a flight of stairs
A flight of steps is a series of steps between floors, between a floor and a landing, or between landings.

Each step comprises a *tread,* which is the upper surface on which the foot is placed, and the *riser*, which is the vertical face of the step (fig. 14.1). It is usual for the tread to project beyond the riser, and this projection is called the

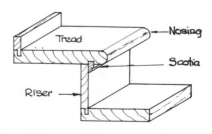

Fig. 14.1 Treads and risers

nosing. In older stairs the joint between the riser and the tread, under the nosing, was sometimes masked by a *scotia* or *scotia mould*.

The treads and risers are supported on either side by timber members known as *strings* or *stringers*, which run the full length of the flight. A string which is fixed to the wall is known as the wall string, and the outer string is one which is remote from the wall. The string supports the treads and risers either by housing the ends of the steps in the string, in which case it is known as a *close string*, or by the string being cut on its upper surface to the profile of the steps, in which case it is known as a *cut* or *open string* (fig. 14.2).

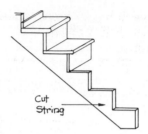

Fig. 14.2 Cut string

The *handrail* is an inclined rail, usually moulded, which serves both as an aid to climbing or descending the flight and as a guard rail above the outer string. The ends of the handrails and strings are housed in stout upright members known as *newels* or *newel posts*. The smaller vertical bars between the outer string and the handrail are known as *balusters*, while a *balustrade* is the infill between handrail and string.

The *stairway* or *stairwell* is the space in which the stairs are contained.

14.3 **Stair dimensions**
The ease with which a staircase may be ascended is dependent upon the amount of effort which a person must put into the action of climbing, and the steps must therefore not be too high, since the effort required to lift the foot increases with the height it must be raised. Similarly, shallow steps, while in

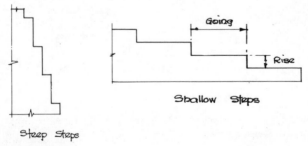

Fig. 14.3 Steep and shallow steps

themselves easier to climb, are more laborious in the long term, since more steps are required to achieve a given height (fig. 14.3).

The amount of height gained by climbing one step, known as the *rise*, and the horizontal distance travelled, the *going*, are both critical dimensions in the design of a stair. The Building Regulations lay down a number of rules which must be adhered to in the design of stairs, and there are two clearly defined types of stairway, namely *private* and *common*, each having its own rules (fig. 14.4).

The private stairway is a series of 'steps with straight nosings on plan which forms part of a building and is intended for use solely in connection with one dwelling'.

The common stairway is similar to the private stairway, but 'is intended for use in connection with two or more dwellings'.

Rules for private stairways are:

a) in consecutive flights of stairs, each step shall have the same rise and the same going;

b) the sum of going dimensions plus twice the rise dimensions shall be a maximum of 700 mm and a minimum of 550 mm;

c) the maximum rise shall be 220 mm and the minimum going shall be 220 mm;

d) the pitch of the stairs shall be a maximum of 42°.

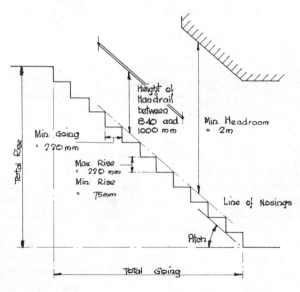

Fig. 14.4 Stair regulations

The foregoing rules are designed to provide a stairway which is comfortable and safe to use. Further safety is achieved by providing suitable handrails and balustrades:

155

e) the stairway should be guarded on each side by suitable screening not less than 840 mm high, measured vertically from the pitch line — the straight line which connects the nosing of each tread in a flight of steps;

f) there should be a handrail provided for any stairway having a total rise of more than 600 mm, i.e. three risers — on one side if the stairway width is less than 1 m, and on both sides if the width is 1 m or more — the height of the handrail being between 840 mm and 1 m, again measured vertically from the pitch line;

g) sides of a landing forming part of a stairway should be protected by suitable screening 900 mm high.

There should also be sufficient headroom safely to descend as well as to ascer the stairs:

h) the clearance should be not less than 2.0 m measured vertically above the pitch line.

The rules relating to common stairways are the same as those for private stairways with the exception of (c), (d), and (g) above. In these cases,

j) the maximum rise shall be 190 mm and the minimum going 230 mm;

k) the pitch of the stairs shall be a maximum of $38°$;

l) the height of screening to landings shall be 1.1 m.

In addition it is desirable to have suitable rest points at intervals when climbing stairs; therefore

m) there shall be no more than sixteen risers in any flight of steps.

14.4 Construction of straight-flight stairs

A straight flight of stairs is usually manufactured in a complete unit off site, only requiring fitting in its final position. The recommended methods of construction and the sizes of members are given in BS 585:1972, 'Wood stairs'. Basic components are treads, risers, and strings. The more common form of construction is for the treads and risers to be housed into the strings, the housing being not less than 12 mm deep and tapered on the underside to allow wedges to be driven in order to ensure a tight joint on the face between the tread or riser and the string. The strings should be a minimum depth of 225 mm, while their thickness may vary between 27 and 50 mm.

The thickness of the treads and risers will depend upon the type and width of the stairway and, in the case of the riser, the type of material used. There are several methods of forming the joint between treads and risers, namely tongue and groove, full housing, and butt or rebated joint (fig. 14.5), and, in order to strengthen the joint between the tread and both the riser and the string, single blocks, not less than 75 mm long and 38 mm wide, should be placed in the internal angles, while the joint between the tread and the bottom of the riser should be strengthened by screws. In addition to strengthening the staircase, the glued blocks also reduce the likelihood of small movements and creaking taking place. It is also recommended that the tongue on the riser should be on the inside face, in order that the nosing is not unduly weakened. The nosing should also be rounded.

Where the width of the stairway exceeds 900 mm, it is recommended that a carriage is placed centrally under the steps to provide intermediate support

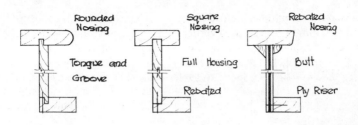

Fig. 14.5 Jointing of treads and risers

to the treads. The actual support is provided by rough brackets which are fixed on alternate sides of the carriage, and the carriage is fixed at the lower end by means of a birdsmouth to a batten on the floor, while at the top end it is fixed to the trimmer (fig. 14.6).

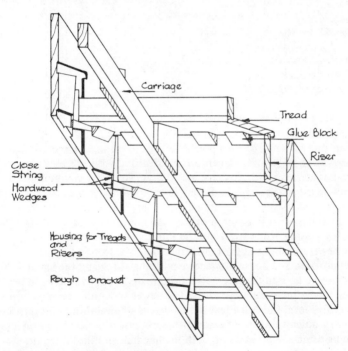

Fig. 14.6 Underneath view of staircase

The wall string is cut at its upper end in such a way that it bears on to the trimmer and also matches up with the skirting (fig. 14.7). The lower end rests on the floor and also matches the skirting, while the top of the string is rebated to receive the plaster wall finish. The whole string should be securely fixed along its length to the wall. The outer string is housed at both ends into the

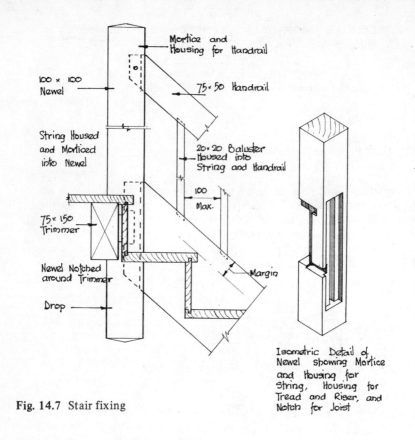

Fig. 14.7 Stair fixing

Isometric Detail of Newel showing Mortice and Housing for String, Housing for Tread and Riser, and Notch for Joist

newel, and there should be a gap of some 50 mm, known as the margin, between the housing of the tread and the top edge of the string.

The newel should be at least 5625 mm² in cross section (75 mm x 75 mm) and be morticed and housed to receive both the string and the handrail. It should also be housed to receive the appropriate treads and riser and be notched for fixing to the floor joists. The newel may project a small distance below ceiling level to form a feature, known as a newel drop, or it may project all the way down to the floor below, in which case it is called a storey newel.

The balusters are housed into both the handrail and the string, or, in the case of a cut string, they are housed into the treads of the stairs. The regulations now require there to be a gap of not more than 100 mm between each baluster if the stairs are likely to be used by persons under the age of 5 years.

An alternative to the vertical appearance created by the balusters is intermediate rails running parallel to the string and handrail. These rails should be housed into or fixed to the newels. It is also advisable to provide extra rigidity by supporting the rails from balusters at intervals along the flight. The maximum 100 mm gap relating to 5-year olds must still be observed.

In the case of a close outer string, it may be easier and cheaper to construct a balustrade using softwood framing and plywood-panel infill (fig. 14.8).

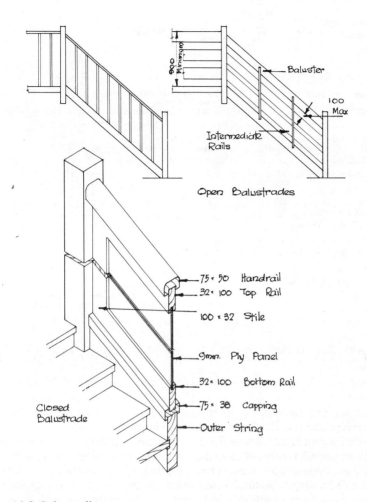

900 Minimum

Baluster

100 Max

Intermediate Rails

Open Balustrades

75 × 50 Handrail
32 × 100 Top Rail
100 × 32 Stile
9mm. Ply Panel
32 × 100 Bottom Rail
75 × 38 Capping
Outer String

Closed Balustrade

Fig. 14.8 Balustrading

15   Principal services

Acknowledgement is due to the Technician Education Council for permission to use the content of the TEC units in this chapter. The council reserves the right to amend the content of its units at any time.

15.1 Services required

The quality of life in most industrial countries requires the provision of many services to all types of properties. The basic services provided, generally by national, regional, or local authorities, are water, gas, electricity, telephone, and drainage.

Water

Water is essential to maintain life in its many forms, and the use to which water is put varies from drinking to washing, from concrete mixing to power generation, and the quality of water required for each specific purpose will vary accordingly. However, the general criterion is 'if it is fit to drink, it is fit to use'. The Water Act 1945 requires the water undertaking to provide a sufficient and wholesome supply. This necessitates that water collected in reservoirs, abstracted from rivers, etc. is treated before it is distributed through a network of pipes to the individual consumers, the treatment usually comprising filtration, sterilisation, and, in certain areas, softening.

The design of the distribution network is governed by the quantity of water and the pressure required at any given location. The quantity is calculated using the actual and anticipated future demands of

a) trade and domestic consumers,
b) fire brigades,
c) energy producers.

The pressure required is determined by

d) the operation of items of plumbing equipment,
e) the supply to tall buildings,
f) the fighting of fires.

Pressure is provided either by gravity feed from storage reservoirs or from pumps, and is expressed in terms of head (i.e. the height to which the water would rise in a vertical pipe connected to the main). This pressure should not generally exceed 70 m head, in order to restrict wastage and keep pipe strengths within economic bounds, but be greater than 30 m head to give sufficient pressure for fire-fighting purposes (fig. 15.1).

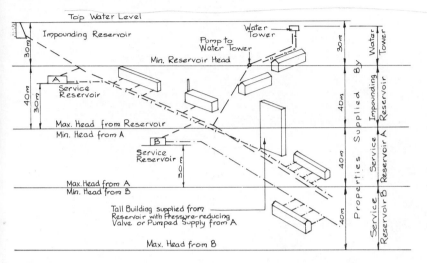

Fig. 15.1 Water distribution (pressure)

The volume of water delivered at a given point is largely dependent on the diameter of the pipe and the pressure head in that pipe (fig. 15.2).

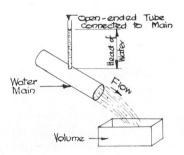

Fig. 15.2 Quantity of discharge

The siting of the distribution network (fig. 15.3) usually follows the roads, thereby enabling the majority of properties to be conveniently supplied, and the supply to a property is provided by a service pipe connected to the distribution main. That portion of the service pipe between the main and the

161

4

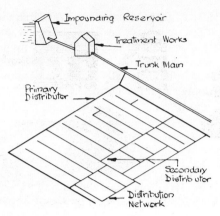

Fig. 15.3 Water distribution (mains)

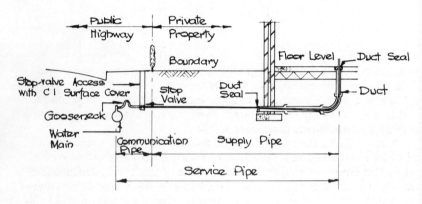

Fig. 15.4 Water-service layout

boundary of the property is known as the 'communication pipe', and the remaining portion laid on the consumer's premises is the 'supply pipe' (fig. 15.4).

Gas

Gas is used mainly as a means of heating and refrigeration, and to a lesser extent for lighting.

Town gas, so called to differentiate it from other types of gas, had traditionally been extracted from coal by a carbonisation process (fig. 15.5). This process of heating the coal to release the gas, along with tar and other chemicals, leaves a residue known as coke. However, the process required large plants, a large labour force, and high-quality coal, all of which made town gas an expensive fuel. Research showed that gas could be produced from crude

162

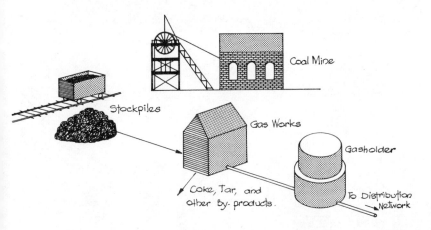

Fig. 15.5 Gas supply

oil, and nowadays this process is used to produce butane and propane gas, which can be conveniently stored under pressure in a liquid form in containers.

The discovery of natural gas in large quantities under the North Sea and other areas has meant that, in Britain, there has been almost a total conversion from town (coal) gas to natural-gas supplies.

Natural gas, having a similar relative density to town gas, but twice the calorific value, could be supplied through the same distribution network of pipes that previously carried the town gas, the network being designed in a similar manner to that of the water-distribution network. The distribution of the gas is in the hands of the area gas boards, under the direction of the Gas Council.

Electricity

The majority of properties in Britain are now supplied with electricity, the uses to which it is put varying from heating, cooling, and lighting, to communications and travel.

Electricity is produced by rotors driving generating equipment. The rotors are driven by steam at pressure, the steam being produced from water heated by gas, coal, fuel oil, or nuclear reactor. Water pressure may also be used to drive the rotors, but only a small amount of electricity is produced in this manner in Great Britain.

From the generator, the electricity is distributed to the individual consumer by a national grid, i.e. a network of overhead cables carrying the electricity at very high voltage (electrical pressure), and regional grids of both overhead and underground cables (fig. 15.6). The generation and primary distribution is the responsibility of the Central Electricity Generating Board (CEGB), while the regional and local distribution is in the hands of the regional electricity boards.

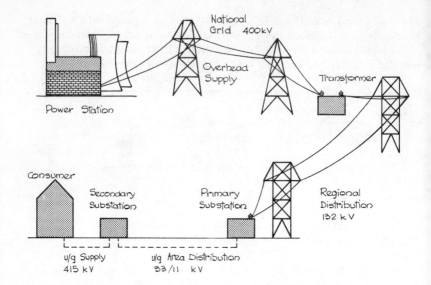

Fig. 15.6 Electricity supply

The regional boards are obliged to provide a supply of electricity to anybody who applies and is prepared to contribute to the installation costs. For new supplies, a board will need to know the approximate load (quantity of electricity) and when it will be required.

It is necessary to reduce the main network pressure (400 kV, 275 kV, 132 kV) to something more usable (33 kV, 11 kV, and 415 or 240 V, depending upon the load to be supplied), and this is achieved by the use of transformers

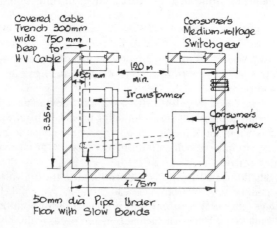

Fig. 15.7 Layout of typical substation

situated at substations (fig. 15.7). These substations are required for all large supplies such as large factories, commercial developments, and housing estates, and, when planning a development, provision must therefore be made not only for the individual supply but also for substation accommodation.

In certain specialist instances, such as hospitals, cinemas, and factories, it may be necessary to provide a standby supply in case of a breakdown in the main supply. This can be provided by
a) secondary supply cables from the public supply,
b) diesel, petrol, or gas engines driving alternators or generators,
c) battery systems.

The decision as to whether a standby supply is required or not is made on the basis of
a) the effect of failure of the main supply,
b) the reliability of the public supply in the area,
c) the cost of providing and maintaining an alternative supply.

As with water, electricity is required in the construction of buildings and it is generally economical to make early provision for the permanent supply to be partially installed and then used in the construction process, rather than having a temporary supply installed to be later followed by the permanent installation.

Underground-cable distribution systems are usually more expensive than overhead systems, but there is less likelihood of damage. These cables may be laid in ducts or conduit, or directly in the ground, depending on the load, quality of insulation, mechanical protection, accessibility, and maintenance requirements. The depths at which cables are laid should not be less than 460 mm below the surface, depending on the position of other services along the route.

Telephone
The telephone is an instrument which is being more widely used every year although, while it is an essential service to the business user, it can still be classed as a luxury for the majority of domestic users. There are two main types of telephone:
a) those connected to the national and international networks,
b) those used only for the purpose of internal communcations.

In many instances the internal system may be linked to the external system via a switchboard.

The system operates by means of two small cables carrying electrical impulses, these impulses being created by sound waves impinging on the mouthpiece or microphone. On arrival at their destination, these electrical impulses create sound waves again from the speaker or earpiece (fig. 15.8). Because the system operates at very low power, the cables are small and many of them can be grouped together to form a larger cable which is generally laid underground. However, in rural areas it is cheaper to carry individual cables above ground, supported at intervals on telegraph poles (fig. 15.9).

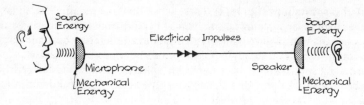

Fig. 15.8 Principle of the telephone

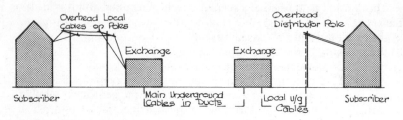

Fig. 15.9 Telephone linking

Drainage

Water arrives at a building by two methods, the first being the natural method in the form of rain, the second by transportation, usually in the form of a piped supply (previously described).

In the first case the property owner has little use for water in that form and wishes to dispose of it quickly. In the second case, having used the water, it is seldom of any further use and its disposal is required.

The means of disposal from the property is known as drainage, and is effected by a system of pipes using gravitational forces to transfer the un-wanted water to some discharge point (fig. 15.10).

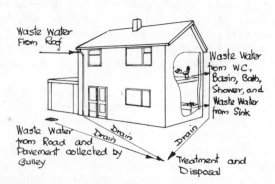

Fig. 15.10 Drainage from the envelope

166

15.2 Substructure provision for services

Each of the services must be brought into, or taken away from, a property in such a manner that neither the structural foundation nor the service itself is detrimentally affected. With the general exception of drainage, provision is made for the entry of the service by means of ducts. This allows construction to take place without the services being present, these being inserted through the ducts at a later stage of the work. The duct must, therefore, be of a suitable size, clear bore, and have no sharp bends which would obstruct the feeding of the service through the duct at a later stage.

Since the excavation for the foundation is usually backfilled as erection of the superstructure proceeds, it is good practice to seal and mark the end of the duct outside the building and to backfill the service-duct area with material which can be easily re-excavated at the appropriate time. To prevent movement or distortion of the duct during subsequent substructure work, weak concrete is frequently placed around the duct (fig. 15.11).

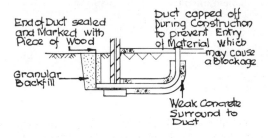

Fig. 15.11 Duct protection

In many cases it is convenient to provide a terminal box in the slab or raft foundation, where two or more services may surface (fig. 15.12).

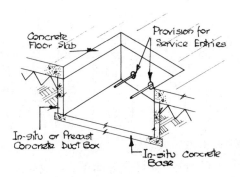

Fig. 15.12 Provision of combined service entry

167

Water

The depth of the water-supply pipe being similar to that of the strip foundation, i.e. approximately 750 mm below ground level, it is usual to provide a
duct immediately above or below the concrete strip foundation, the duct
terminating at ground-floor level. A change of direction from horizontal to
vertical plane is effected by bends of suitable radius such that the supply pipe
will alter direction without kinking or large deformation.

It is good practice to seal the ends of the duct, in order to prevent
a) the entry and breeding of vermin,
b) any circulation of cool air which might tend to freeze the service,
c) the entry of ground water which may have corrosive effects on the pipe
   (see fig. 15.4).

Gas

The entry of the gas service pipe into a building may be either underground
or above ground, depending on the situation and the gas region's requirements
(figs 15.13 to 15.15).

The minimum depth of the gas service is 375 mm.

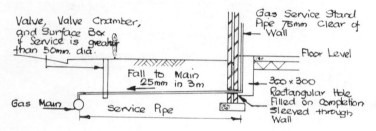

Fig. 15.13 Gas-service layout

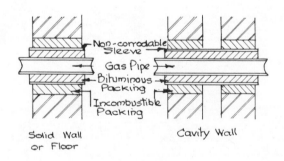

Fig. 15.14 Gas ducting through walls (N.B. Steel pipe should have min.
12 mm bituminous packing all round.)

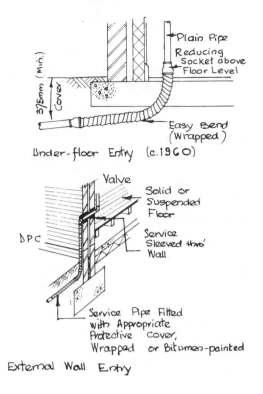

Plain Pipe
Reducing Socket above Floor Level
Easy Bend (Wrapped)

375mm (Min.) Cover

Under-floor Entry (c.1960)

Valve
Solid or Suspended Floor
DPC
Service Sleeved thro' Wall
Service Pipe Fitted with Appropriate Protective cover, Wrapped or Bitumen-painted

External Wall Entry

Fig. 15.15 Gas-service entry

Electricity

The electricity supply cable to the property is housed in a duct as it enters the substructure (fig. 15.16). This duct may be of 100 mm diameter earthenware pipe having watertight joints or of 32 mm plastics (p.v.c.) ducting. The size of duct depends on the size of the supply cable, the radius of any duct bend being greater than twelve times the diameter of the cable passing through that duct. Once the cable is inserted in the duct, the outer end of the duct is sealed using a waterproof mastic seal to prevent ingress of water to the duct.

The modern method of electrical service installation is to use a looped supply (fig. 15.17). This is laying one service cable to alternate houses and looping a shorter length to the intermediate houses; therefore two ducts will be required in alternate properties — one for the service cable and one for the loop to the next house.

In certain areas of the country where ground subsidence is a problem, supplies may be taken overhead, in which case the service cable is attached to the building on insulated brackets and the supply is taken to the inside of the envelope through a p.v.c. duct in the wall above ground level.

2

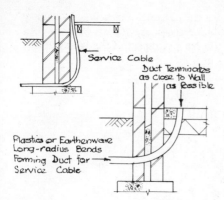

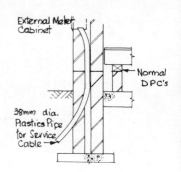

Fig. 15.16 Electricity-service entry

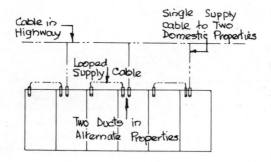

Fig. 15.17 Plan of looped electricity supply

Telephone

Since the telephone cable is small and pliable, the wires will pass through a hole drilled in a timber window frame or external wall, suitably sealed after insertion to prevent moisture penetration. However, a duct may be erected through the substructure in the form of a small-diameter pliable pipe (such as the alkathene pipe used for water-supply pipework), again being suitably sealed to prevent moisture penetration.

Drainage

It is essential that the drain in any substructure work will at all times remain watertight and undistorted by any settlement. The drain is therefore surrounded in concrete wherever it is under a building, and extra-strength pipes, such as cast iron, may be required where heavy loads are applied to the foundation. The drain is terminated with a socket just above floor level, so that jointing of pipes to it from above is facilitated.

Access to a drain under a building for the purpose of clearing blockages is usually impractical, and for this and other reasons any vertical bends should be of long radius and well supported to take the impact of the falling water (fig. 15.18).

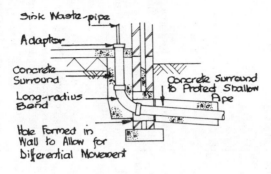

Fig. 15.18 Outlet for internal drainage

Where access to a bend is reasonably close, as is commonly the case with the discharge from a kitchen sink into a back inlet gulley, a sharp bend may be used, again with the proviso that it is adequately supported. The drainage above ground from the sink, being of smaller-diameter pipework than that below, will require some form of adaptor, usually at ground level.

15.3 Performance characteristics of service-installation materials
The performance characteristics of the basic materials used in service installations are given in schedules – see figs 15.19(a) and (b).

15.4 Protection of service installations
Services must be protected from both physical and climatic damage, since it is likely that more than one consumer may be affected. The cost of the damage may not only be that of a repair bill but also that of lost production, or in extreme cases loss of life.

Water
The water supply system has generally no protection other than that of depth. That depth, 900 mm minimum for a main and 760 mm minimum for a service pipe, is below that to which a garden is normally dug and, more important, it is below the depth to which a normal winter's frost will penetrate the ground, thereby avoiding frozen supplies and the consequent fractured pipes.

Ground movement may be allowed for by using semiflexible pipes and jointing techniques, which will move with the ground, or strong rigid pipelines which will span soft spots. Pipes may also be designed to allow movement, such as the goose-neck on a service pipe (see fig. 15.4).

In corrosive soils, the pipe may be protected from ground acids by wrapping it with a suitable tape or by coating it with a bituminous compound.

Supply: Water			
	Situation		
Material	*Main supply*	*Services*	*Duct*
Lead	Not used	Old-fashioned. Flexible, weak but durable. More expensive than other materials. Difficult to joint. Heavy. Not attacked by most soils.	Not used
Copper	Not used	Ductile but with good strength. Easy to joint. Attacked by strong acids but widely used.	Not used
Alkathene	Not used	Flexible plastic with good pressure resistance. Easily cut. Not materially affected by ground conditions.	Used as a sleeve for copper service. Easily built in, but care must be taken to avoid 'kinking'.
Cast iron	Widely used. Good pressure resistance. Strong. Various joints available. Attacked by acidic soil so should have bituminous coating.	Occasionally found in galvanised form. Good external corrosion resistance although subject to internal corrosion.	Strong. Suitable where foundations are heavily loaded.
Steel	Strong but subject to corrosion; therefore requires expensive protection coat. Suitable for larger diameters.	Seldom used	Can be used for lightly loaded foundations, especially in flexible form, but must be wrapped for corrosion protection.
Concrete	Good load-bearing in larger diameters. Easy to joint but heavy to manoeuvre. Good corrosion resistance to most soils.	Not used	Used to form openings in walls for ducts, in either precast or in-situ form.

Material	Main supply	Services	Duct
P.V.C.	Used economically in rural areas where large diameters are not required. Easy to joint. Poor load-bearing capacity, but flexible.	Not used	Widely used but should be surrounded with concrete in heavily loaded areas.
Pitch fibre and asbestos cement	As for p.v.c. but not as flexible.	Not used	As for p.v.c.
Earthenware	Not used	Not used	As for p.v.c.

Supply: Gas

Material	Main supply	Services	Duct
Copper	Not used	Used with plastics protective outer skin in most types of ground.	Not used
Lead	Not used	As for water. Pipe should be of gas weight to BS 602.	Not used
Steel (mild)	As for water.	Most common. Rigid screwed joint in mild steel or malleable cast iron. Corrosion protection required in most soils. Used especially where service passes through duct, and wrapped 150 mm beyond each end of sleeve.	Frequently used for sleeving, but should have bituminous coating.
Cast iron	As for water.	Not used	As for water.
Pitch fibre	Not used	Not used	As for water.
Earthenware	Not used	Not used	As for water.

Fig 15.19 (a) Performance characteristics of service – installation materials

Supply: Electricity

Material	Main supply	Situation	
		Services	Duct
Copper	Ideal conductor, provided adequate insulation is used. Easily jointed.	As for main supply.	Not used
Aluminium	Not as good a conductor as copper but requires less insulation and is suitable only for low and medium voltages.	As for main supply.	
Plastics	Good insulation material, but easily cut.	As for main supply.	Ideal since provides good insulation. Should have concrete protection.
Earthenware	Good insulation, but fragile.	As for main supply.	As above but requires concrete protection only in heavily loaded areas.

Supply: Telephone

Material	Main supply	Situation	
		Services	Duct
	As for electricity		

174

Supply: Drainage		Situation	
		Situation	
Material	Main supply	Services	Duct
Earthenware	Widely used. Good resistance to chemical attack. Easily jointed and laid. Requires support under heavy loads.	As for main supply.	
Concrete	As for earthenware, but used only for large diameters and may be attacked by sulphates.		
Cast iron	Good corrosion resistance but should have bituminous coating. Used in heavily loaded areas.	As for main supply.	
Pitch fibre	Flexible. Good corrosion resistance. Easily jointed. Requires adequate support. Not manufactured in large diameters.	As for main supply.	
P.V.C.	Lightweight. Easily jointed and laid. Brittle at low temperatures, but generally flexible. Requires adequate support. Not manufactured in large diameters.	As for main supply.	
Asbestos cement	As for pitch fibre, especially if coated with bitumen, otherwise liable to sulphate attack.	As for main supply.	

Fig. 15.19(b) Performance characteristics of service-installation materials (cont'd)

Gas

The amount of protection against corrosion required for gas mains and services will depend on the nature of the soil. Cast-iron pipes have only a bituminous external-surface coating, since they have good corrosion resistance. Steel and wrought-iron pipes should have a hessian or bituminous wrapping covered with asphalt, and exposed pipes should have a red-lead or bituminous paint coating. No other form of mechanical protection is required save that of depth in the ground, which should be approximately 750 mm.

Electricity

Electricity supply cables, being laid at a lesser depth (minimum 460 mm below ground level) than the gas and water installations, are more prone to physical damage when any excavation is begun. It used to be the practice to cover the cable laid directly in the ground with a precast concrete slab similar to a coping (fig. 15.20). Since more recent times, cables laid directly in the ground have been armoured with metal sheathing or wire armour (fig. 15.21). Where cables are drawn into ducts the cables need no armour, since the ducts provide sufficient mechanical protection.

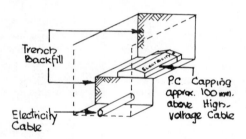

Fig. 15.20 Cable protection

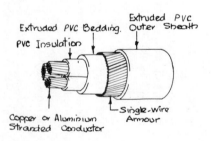

Fig. 15.21 Armoured cable for laying direct in the ground

Being underground, the cables have to be protected against moisture to prevent short circuits being formed, or corrosion of the sheathing by ground

176

water and chemicals. This protection is obtained by serving the cable with bituminised jute or hessian tape, or an extruded p.v.c. outer sheath.

Cables should be protected from fractured water, gas, petrol, oil, and drainage pipes by good route-planning. Care should also be taken to see that the electricity cables are not in close proximity to the telephone cables, since the magnetic effect of the electrical current may affect the operation of the telephone communication system.

Temperature will have little effect on the cable, unless it is excessive, provided that the cable is laid slack with allowance for expansion and contraction of the metallic conductor. Overhead cables are at such a height as to be out of reach from the ground but below normal aircraft flying heights.

Telephones

The early telephone lines were carried overhead, sometimes up to 30 m above ground so as to pass over tall buildings, and still are on housing estates where an underground cable terminates on a distribution pole so positioned that wires radiate from the pole to serve a number of properties. The cables still have to be high enough to allow normal traffic to pass underneath, but the problem with this method of supply is the damage likely to be caused by high winds and the extra weight of snow and ice on the wires. These problems are overcome by laying the wires in cable ducts, the pair of wires for an installation being grouped with others to form a larger cable, covered with a plastics sheath to keep moisture from the wires. The earthenware ducts are sometimes able to carry up to six cables and are sufficiently strong to prevent accidental damage by excavation. As with electricity cables, temperature changes will not affect the cables provided that there is 'slack' in the system.

Drainage

The drain, being laid to falls, is usually sufficiently deep to avoid most physical and climatic damage; however, as with the water supply, the pipework must be

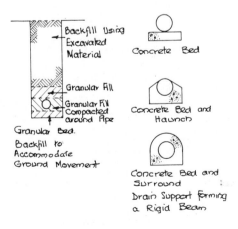

Fig. 15.22 Drainage protection

designed to accommodate ground movement, either by use of flexible joints in conjunction with a flexible bedding material or by forming a rigid beam. There is a sufficient range of pipe materials to take care of most external chemical attacks. The main danger to drainage pipeworks is at or near surface connections where the pipes are shallow. In this case the pipes should be surrounded with concrete for at least 600 mm below the surface (fig. 15.22).

On new estates it is convenient to provide one excavation for as many services as possible (fig. 15.23). This method has two main advantages:
a) economy of laying in one trench,
b) less likelihood of damage by mechanical means.

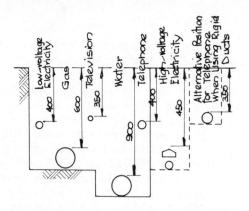

Fig. 15.23 Layout of services in common user trench

16   Drainage

Understands the basic principles of drainage installations.

16.1   *Identifies main types of effluent and describes their characteristics.*
16.2   *Describes the general principles for the collection and removal of surface water from roofs and paved areas.*
16.3   *Sketches and describes typical eaves, gutters, rain-water pipes and standard fittings including jointing and method of fixing.*
16.4   *States the general principles for the construction of an underground drainage system.*
16.5   *Sketches and describes typical drainage systems.*

Acknowledgement is due to the Technician Education Council for permission to use the content of the TEC units in this chapter. The council reserves the right to amend the content of its units at any time.

16.1 Types of effluent

Effluent or sewage can be defined as unwanted water which has to be disposed of. Sewage is water-borne human, domestic, trade, and farm waste and may also be surface or subsurface water. The majority of sewage must be treated before it can be discharged into a watercourse, the ground, or the sea.

Sewage is a complex substance, the composition of which varies greatly depending on the quantities of the various wastes which are discharged into the underground drainage system. It consists of *solids, chemicals, bacteria*, and *colloids* — particles of a jelly-like character, which are inbetween matter in suspension and matter in solution. The drainage system must be resistant to the scouring action of some of the solids, chemical attack, and the action of bacteria.

Bacteria are minute vegetable growths which multiply rapidly; they need moisture but are affected by cold. There are two types: *parasitic*, which need a living host, and *saprophytic*, which live on dead matter. The latter is the more important in sewage disposal, and again there are two types: *anaerobic*, which live without air, and *aerobic*, which require air. Crude sewage has a large amount of anaerobic bacteria but few aerobic, and at the sewage-treatment works the aerobic bacteria are increased and the anaerobic reduced by aeration. The anaerobic bacteria cause the sewage to decompose, which results in the solids being liquified and gasified, and the aerobic bacteria allow oxygen to combine with the organic matter, thus forming stable harmless compounds such as nitrates.

16.2 Collection and removal of surface water

Surface water occurs in one of two ways: either it comes from the skies in the form of rain, sleet, or snow — usually known as rain-water — or is natural water rising from the ground in the form of springs.

In whichever form it occurs, it is unwanted water as far as houses and paved areas are concerned, since in the first instance it creates extra loads on the roofs and in the second it is inconvenient both to pedestrians and to vehicular traffic. It is therefore essential to our modern way of life that this water be removed quickly and disposed of at some more convenient place.

Both pitched and 'flat' roofs have slopes which cause the rain-water falling on them to flow over their surfaces by gravity, to convenient collection points. The initial collection point for the pitched roof is at the eaves, where the roof covering discharges the rain-water into a *gutter* (fig. 16.1). The gutter runs the

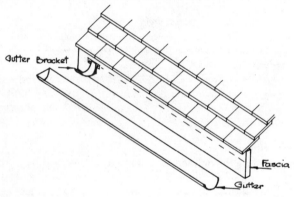

Fig. 16.1 Gutter at eaves

full length of the eaves and is generally supported by means of brackets which are fixed to the fascia board. It can be fixed level or have a small amount of fall (1:150 to 1:600) to the point of discharge. The positioning of these outlets or nozzles depends on the area of roof being drained (fig. 16.2), the anticipated rainfall in a given period of time, and the size of the gutter itself. The outlet discharges either directly or by means of a *rain-water head* or *hopper* into a vertical pipe known as the *rain-water pipe* or *downpipe* which at its lower end is terminated by a *rain-water shoe* which discharges the rain-water into a surface water drain or, by a direct connection, into a back inlet gulley. Methods of calculating the sizes of gutters and downpipes are given in BRE Digest 188.

Paved areas are also laid to falls so that the surface water is directed to suitable collection points in the form of gulleys. These gulleys, situated at suitable intervals, discharge the water into the underground drainage system.

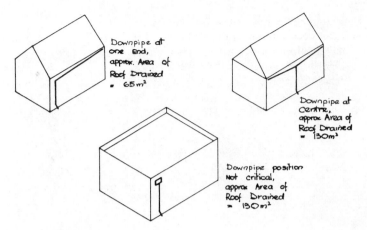

Fig. 16.2 Roof drainage

16.3 Rain-water gutters and pipes

The Building Regulations (clauses N8 and N9) require gutters and rain-water pipes to be adequately sized and supported without restricting thermal movement, composed of suitable strong and durable materials, and appropriately jointed so as to remain watertight and not cause dampness in the building.

There are several shapes of gutter, see fig. 16.3. The rain-water pipes (r.w.p.) are generally circular but rectangular ones are available, usually for use with box-section gutters. Where there are overhanging eaves, a swan-neck is required to allow the r.w.p. to be fixed to the wall.

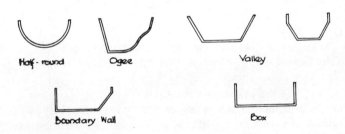

Fig. 16.3 Gutter shapes

Gutters, r.w.p.'s and their special fittings (fig. 16.4) are manufactured in a variety of materials — aluminium, asbestos cement, cast iron, galvanised steel, precast concrete, p.v.c., and zinc — and each material requires its own method of jointing and support.

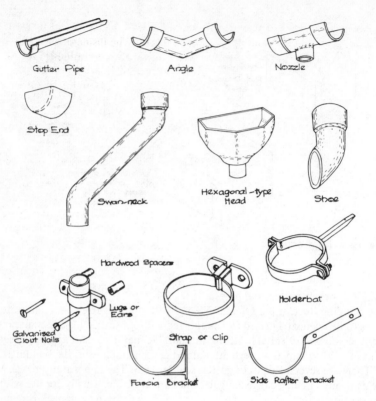

Fig. 16.4 Rain-water fittings

Gutters

Where gutters have socket-and-spigot joints, the socket should face the direction of the flow of water — this is also the case with all other drainage pipework. The gutter can be fixed by brackets to the fascia, wall, or rafters.

Gutters are available in the following materials.

a) *Aluminium*   Formed by pressing, casting, or extrusion, they require little maintenance and are jointed with bituminous mastic, requiring support at 1.8 m intervals.

b) *Asbestos cement*   Formed by a spray process, they do not corrode or need painting and are jointed by screws and bitumen or joint pads and require support at 1 m intervals.

c) *Cast iron*   These are heavy and require regular painting both inside and outside to prevent rust. They are jointed by bolts with red lead or lead wool and require support at 900 mm intervals.

d) *Galvanised steel*   These do not have a long life, are lighter than cast iron, but are jointed and supported in a similar manner.

e) *Precast concrete* These are constructed as an integral part of the eaves and are jointed with suitable mortar or mastic. The inside surface may be lined with suitable non-ferrous metal or may have a bituminous coating.

f) *P.V.C.* Lightweight, easy to fix, do not require painting nor do they corrode, but will not support a ladder. Sunlight can cause slight colour change and distortion. The sections are clipped together and require frequent support.

g) *Zinc* Similar to aluminium, but lighter in weight, they require support at 750 mm centres.

Rain-water pipes

Pipes are fixed to the wall by straps, holderbats, or galvanised screws or nails driven through lugs or ears cast on to the socket end of the pipe (loose ears are also available). The pipe should be clear of the wall by approximately 40 mm to allow for maintenance and painting. In the case of the strap and holderbat fixings, the clearance is predetermined, but special hardwood bobbins or distance pieces are required for lug fixings. Standard domestic sizes are either 63 or 75 mm diameter and, in order to prevent wind-blown debris from entering the rain-water pipe and causing a blockage, it is advisable to insert a galvanised steel-wire balloon grating at the gutter outlet.

Rain-water pipes are available in the following materials.

a) *Aluminium* Fixed by ears. Jointing may be loose or caulked.

b) *Asbestos cement* Fixed by straps or holderbats. Jointing may be with a 1:2 cement:sand mix or loose.

c) *Cast iron* Fixed by ears. The joints may be made with lead wool, cement:sand mix, or loose.

d) *Galvanised steel* Fixed by ears. The joints are left loose since this type is suitable only for temporary work.

e) *P.V.C.* Fixed by straps. Jointing is by rubber 'O' rings.

Rain-water pipes which are situated inside a building must be jointed and, together with the r.w.p., must be able to withstand an internal build-up of pressure which may be caused by a blockage.

16.4 Underground drainage systems

Any drainage system relies for the most part on the flow of effluent, by gravity, from the point of entry to the point of discharge. Pumping is occasionally used but is expensive to install, maintain, and run.

Definitions (fig. 16.5)

a) *Drain* Pipe used for drainage of waste water from a single building or a number of buildings within the same curtilage. It is private property and maintained by the property owner.

b) *Private sewer* Pipe used for drainage of waste water from a particular group of buildings. It is the responsibility of the owners of the buildings concerned.

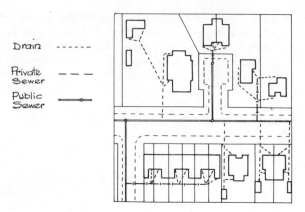

Fig. 16.5 Definitions

c) *Public sewer*   Pipe used for drainage of waste water of several properties, for which the Local Authority is responsible.
d) *Soil drainage*   Drainage of waste water from soil and waste fittings, i.e. water-closets, urinals, sinks, baths, bidets, basins, etc.
e) *Surface-water drainage*   Drainage of waste water from roofs, yards, pavements, etc., consisting of rain-water which does not require cleansing.
f) *Separate sewerage system*   Two separate drainage systems, one carrying foul or soil water to the sewage-purification works, the other carrying rain-water to discharge directly into a watercourse or the sea.
g) *Combined sewerage system*   Waste water from both rain-water and soil fittings is conveyed in one sewer to the sewage-treatment works.
h) *Inspection chamber (i.c.)*   A means of access into a drain for inspection or cleansing purposes – sometimes called a manhole (m.h.).

Principles of external drainage
a) Drains must be airtight, watertight, and have a clean bore and an even invert.
b) They must have a solid foundation, an adequate and uniform gradient, and be laid in straight lines between points of direction change.
c) Adequate means of inspection and cleansing must be provided.
d) Inspection chambers must be provided at changes of direction and in convenient positions to receive the maximum number of branches.
e) The maximum distance between i.c.'s is 90 m, and in no case should any point on a drainage scheme be more than 45 m from an i.c. In the case of a sewer or drain connection, if there is no manhole or i.c. on the sewer or drain at the point of connection, there must be an i.c. on the connecting drain or branch not more than 12.5 m from the connection (fig. 16.6).
f) An i.c. or rodding eye (see fig. 16.8) must be provided at the highest point of a private sewer.
g) Drains and private sewers must be of such a size and laid at such a gradient as to be self-cleansing and able to carry away the maximum volume of

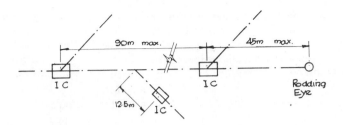

Fig. 16.6 I.C. positions

material discharged into them. In no case should the soil drain be less than 100 mm internal diameter or a rain-water drain less than 75 mm internal diameter.

h) At all junctions, the flow from the tributary drain should be obliquely directed in the direction of the flow of the main drain.

j) Where a drain trench passes near a load-bearing part of a building, concrete backfill is required (fig. 16.7).

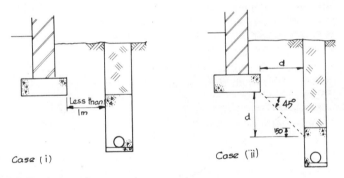

Fig. 16.7 Drain trenches

    Case (i) Trench less than 1 m away — requires concrete backfill to the level of the underside of the foundation.

    Case (ii) Bottom of trench lower than a line drawn down at 45° from the near bottom corner — concrete fill required to a height of 150 mm above this line.

k) Inlets to drains should have a suitable trap, having a seal of not less than 50 mm.

l) Inspection chambers should be of a suitable size and strength.

Notes on principles

a) In order that the drains shall remain watertight under all working conditions, it is necessary to provide either a granular or a concrete bed for the pipes. In cases of poor ground, it is as well to ensure that the pipe joints do not fail, either by haunching or by surrounding the drain with concrete (fig. 15.22), in which case a rigid type of joint may be used. Where reasonable ground is encountered, a flexible type of pipe joint can be used in conjunction with a granular surround.

b) Drains should be laid to such falls as will ensure a flow rapid enough to prevent sedimentation of solid matter which might lead to a blockage. The pipe size and fall must also be chosen to ensure that the pipe does not run full and thereby cause suction which could unseal traps in gulleys or in fittings within the building. A suitable velocity of water flowing in a drain is approximately 0.75 m/s, and CP 301:1971, 'Building drainage', recommends that, at peak flow, pipes should be designed to run at 90% capacity, i.e. with the depth of flow in pipes not exceeding 0.75 of the pipe diameter. General guide-lines for design are

  i) not more than 20 dwellings connected to a 100 mm diameter drain;
  ii) not more than 150 dwellings connected to a 150 mm diameter drain;
  iii) gradients:

Pipe dia.	McGuire's rule	Max. gradient	Min. gradient
100 mm	1:40	1:4	1:80
150 mm	1:60	1:6	1:150
225 mm	1:90	1:10	1:200
300 mm	1:120	1:20	1:250

Gradient may be reduced or increased if a system is designed using hydraulic tables such as 'Crimp and Bruges' or design formulae such as Scobey's.

c) Blockages occur in the best-designed systems and are usually cleared by rodding through the pipeline with flexible cane, metal, or plastics rods to which a suitable ram has been attached. To get the rods into the pipe and to allow for jointing the rods together, a means of access is required. The traditional means of access is the i.c. — usually constructed of brick concrete or precast concrete. Manholes and i.c.'s are, however, expensive to build, and designers are now tending to use fewer i.c.'s and substitute rodding points or eyes where possible. The rodding eye (fig. 16.8) consists of a

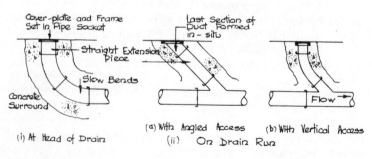

(i) At Head of Drain

(a) With Angled Access   (b) With Vertical Access
(ii) On Drain Run

Fig. 16.8 Rodding eyes

vertical or inclined length of pipe, terminating with a sealing-plate at ground level, while the bottom of it makes a curved junction with the drain, thereby enabling flexible rods to be pushed into the system via the access pipe.

d) When backfilling, care should be taken to see that the first 300 mm on top of the pipe is well consolidated by hand and free from large hard pieces of material which may have a tendency to fracture the pipe. In practice it is usual to use similar material to that used for bedding (i.e. pea gravel). The remaining backfill should be placed, not dropped, on to the trench, thereby reducing the risk of fractures due to impact loading.

Materials

a) *Vitrified clay* (BS 65 & 540)   Supplied in lengths between 0.6 m and 1.5 m. Pipes available with either socket-and-spigot ends or double-spigot ends. Methods of jointing socket-and-spigot pipes are (i) flexible rubber ring seal (flexible joint), (ii) plasticised p.v.c. (flexible joint), (iii) cement and sand (rigid joint), (iv) clay (an old method not now acceptable). Double-spigot pipes are jointed by means of a polypropylene coupling and flexible rubber rings. Clay offers a high resistance to attack from a wide range of chemicals, including sulphates.

b) *Concrete* (BS 556)   Manufactured in diameters from 225 mm upwards, either reinforced or unreinforced. Care should be taken to guard against internal or external sulphate attack. Jointing is by means of flexible rubber ring seals.

c) *Cast or spun iron* (BS 437)   Manufactured in 3 m lengths and upwards, therefore less joints required. Capable of resisting high pressures and until recently required for all drains passing under buildings. Socket and spigot joints are usually made with run lead or lead wool caulked home, to allow slight flexing of the joint.

d) *Pitch fibre* (BS 2760)   Manufactured from wood pulp, cellulose fibre, and bituminous pitch in lengths between 1.7 m and 3.0 m. A flexible material, resistant to chemicals. Pipes have spigot ends of $2°$ taper and are jointed by push-fitting into a double collar of the same material or a polypropylene coupling as for clay pipes.

e) *P.V.C.* (BS 4660). Available in 1 m, 3 m, and 6 m lengths. Lightweight, easily handled in long lengths, more easily damaged in stacking, rodding, or impact loading. Becomes brittle at low temperatures. A complete range of fittings is now available, including i.c.'s (Marley and others). Jointing in the case of socket-and-spigot or double-spigot pipes is by means of rubber-ring-insert or solvent-cement joints.

f) *Asbestos cement* (BS 3656)   Manufactured in lengths of 2.5 m. Substances that attack concrete may also attack a.c. Bitumen-dipped pipes are available, the coating giving increased resistance to attack both internally and externally. The method of jointing spigots and sockets is with rubber rings.

g) *Types of joints*   See fig. 16.9.

h) *Drainage fittings*   See fig. 16.10.

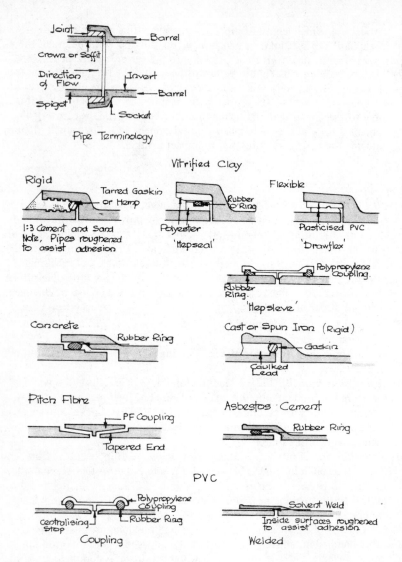

Fig. 16.9 Pipe joints

Tests on drains

The Building Regulations 1976 require that 'any drain or private sewer shall be capable of withstanding a suitable test for watertightness after the work of laying has been carried out.' The work is deemed to include any haunching surrounding or backfilling. It is, however, usual to test the drainage scheme immediately after laying – this checks that the pipes themselves, together with fittings and joints, are watertight before any further work is carried out.

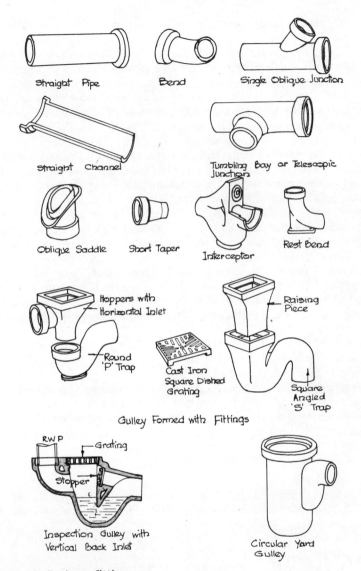

Straight Pipe

Bend

Single Oblique Junction

Straight Channel

Tumbling Bay or Telescopic Junction

Oblique Saddle

Short Taper

Interceptor

Rest Bend

Hoppers with Horizontal Inlet

Round 'P' Trap

Cast Iron Square Dished Grating

Raising Piece

Square Angled 'S' Trap

Gulley Formed with Fittings

RWP

Grating

Stopper

Inspection Gulley with Vertical Back Inlet

Circular Yard Gulley

Fig. 16.10 Drainage fittings

Types of test on new drains

a) *Hydraulic or water test*   This test would automatically be applied to a drain if a blockage occurred and is therefore favoured by many local-authority building inspectors. It consists of closing the lower end of a drain run by means of a plug or stopper; filling the drain with water to such a

189

height as will apply an appropriate pressure (1.5 m maximum head); waiting for the level to settle, thereby allowing for absorption of water into the pipe material; and then waiting 30 minutes to see that no more than the permitted drop in level takes place. Care should be taken to see that no air is trapped in the drain.

b) *Air test*　　The system to be tested is firmly plugged at both ends and air is pumped in until the pressure is slightly in excess of 100 mm water-gauge on a manometer connected to the system. After allowing the system to balance, the pressure should not fall from 100 mm to less than 75 mm during a period of 5 minutes without further pumping. This test is used when no water is available on site or when the discharge of the test water would cause inconvenience. The disadvantage of the test is that there is no visual indication of which joint(s) or pipe(s) has failed.

c) *Smoke test*　　Similar to the air test, but smoke is pumped into the pipeline or smoke bombs are inserted to give the pressure increase. This test gives visual aid to the location of faults, since smoke will penetrate pea-gravel bedding material.

d) *Ball test*　　This consists of rolling a ball whose diameter is some 10 mm less than that of the pipeline down the line from the higher end. This tests that (i) there is fall on the pipeline, (ii) there is no blockage (waterproofs, stoppers, etc.), and (iii) there is a clear bore (especially with cement—sand joints).

e) *Mirror test*　　Very occasionally used. Mirrors are placed in the invert or channel of adjacent i.c.'s, and by using a torchlight any bad joints can be seen along the inside of the pipes.

f) *Visual test*　　By inspection, checks on the straightness of the drain run, quality of pipe, type of joint, and type of bedding material can be made.

g) *Colour test*　　Used to trace flows of existing drains and connections by means of bright red, green, or blue fluorescent powder which dyes the water flow.

Inspection chambers

a) *Regulations*　　The Building Regulations require an inspection chamber to be designed and constructed of brickwork, concrete, or other suitable and durable material. The chamber must
   i) be able to sustain the loads imposed upon it by traffic, ground thrust, etc.;
   ii) be adequate in size for working in and carrying out inspection;
   iii) be watertight against both internal and external water pressure;
   iv) have a safe means of access to the drain by means of ladders, step-irons, ramps, etc.;
   v) have a non-ventilating cover of suitable strength and manufactured out of durable material (fig. 16.11);
   vi) where any part of the system in the chamber is in open channel, be provided with benching which has a smooth impervious finish, providing a safe foothold and so formed as to guide the flow towards the main pipe into which the chamber discharges.

190

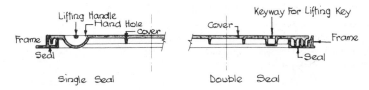

Fig. 16.11 Inspection-cover sections

If the chamber is within a building, the construction should be so designed (cover and frame included) as to withstand the maximum pressure which any blockage in the system would cause in that chamber. The cover should be non-ventilating and fit into the frame with an airtight seal, and be secured to the frame with a set of removable corrosion-resistant bolts.

b) *Size of inspection chambers* Size is governed by the necessity to provide room for a man to work and insert drain rods and stoppers without difficulty. Fixed dimensions are not applicable since the size may depend on the angle of the main drain, the position of the branch drains entering the chamber, and so on.

Guide to sizing of rectangular chambers:

Depth to invert	Min. length	Min. width	Min. height above benching	Concrete base thickness
Up to 0.6 m	0.6 m	0.45 m	–	100 mm
0.6–0.9 m	0.75 m	0.60 m	–	100 mm
0.9–1.8 m	1.00 m	0.70 m	–	150 mm
1.8–4.5 m	1.35 m	0.80 m	2.0 m	230 mm
Over 4.5 m	1.50 m	1.15 m	2.0 m	230–450 mm

In chamber design, provision should also be made for branches. Allow 0.3 m in chamber length for a 100 m branch and 0.375 m for a 150 mm branch, plus adequate allowance on the downstream side for the angle of entry, and 0.3 m in chamber width for each side with branches, plus whichever is the greater of 0.15 m or the diameter of the main drain.

For a chamber 0.8 m deep with two $\phi 100$ branches on one side and one $\phi 150$ branch on both sides, main channel $\phi 225$ (fig. 16.12):

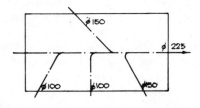

Fig. 16.12 Chamber sizing

length = 2 x 300 mm + 1 x 375 mm + C

$\quad$ = 975 mm + C $\quad$ (say C = 125 mm)

$\quad$ = 1.100 m

width = 300 mm + 300 mm + 225 mm

$\quad$ = 0.825 m

Therefore design the chamber to the nearest unit size above 1.100 m x 0.825 m, i.e. brickwork 1.125 m x 0.900 m internal size.

c) *Typical construction* $\quad$ See figs 16.13 and 16.14.

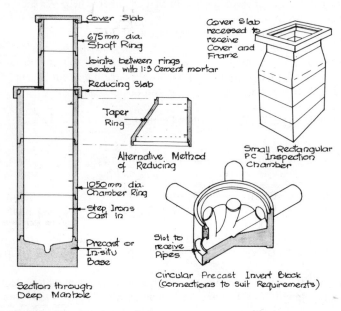

Fig. 16.13 Precast concrete manholes

d) *Backdrop m.h. or i.c.* $\quad$ Excerpt from CP 2005, 'Sewerage': 'Where a sewer connects with another sewer at a materially lower level, a manhole may be built on the lower sewer incorporating a vertical or near vertical drop pipe from the higher sewer to the lower one.'

$\quad$ The drop pipe may be either outside the shaft and encased in concrete or supported on brackets inside the shaft, which should be suitably enlarged. If outside, the higher sewer should be built through the shaft wall to form a rodding eye and inspection access (fig. 16.15). Similar access is necessary if the drop pipe is inside the manhole unless the drop pipe is of cast iron. The drop pipe should terminate at its lower end in a 90° bend properly set

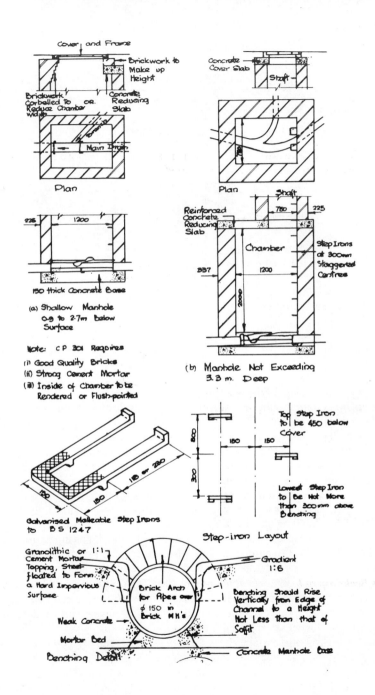

Fig. 16.14 Brick manholes

193

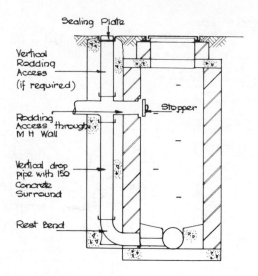

Fig. 16.15 Backdrop inspection chamber

in concrete to turn the flow from the branch in the direction of the main flow.

16.5 Typical drainage systems

A drainage system should be designed so as to provide an efficient, economic, and practical layout. An efficient system is achieved by correct pipe-sizing and falls, thereby achieving self-cleansing velocities. Economy is obtained by careful layout planning. The underground system should make maximum use of inspection chambers for branch-connection purposes, have minimum lengths for drain runs, and where possible follow the contours of the ground to avoid deep trenches. Positioning of the collection points for both foul and surface water must also be considered, since the grouping of these points both inside and outside a building may reduce the need for underground runs.

The practical aspects relate to the positioning of the drains in relation both to the buildings one to another and to other underground services so that the minimum disruption is caused, both during installation of the system and at any future time during the completion of the site works (fig. 16.16).

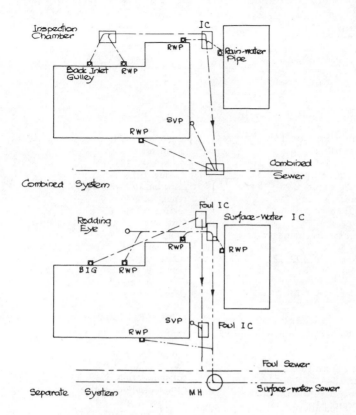

Fig. 16.16 Typical drainage layout

195

Index